THE GIFTING BIRDS

TOWARD AN ART OF HAVING PLACE AND BEING ANIMAL

Charles Jones

1985
DREAM GARDEN PRESS
Salt Lake City

For Molly
We live in all the best places,
The ones unnamed and unnameable.

The author would like to thank:

Harold I. Silverman, Associate Editor of the *San Francisco Examiner*, for publishing six of the chapters in this book in *California Living*.

The editors of *Northwest*, *Westways*, *The Beachcomber*, *Pacific Search*, the *Sierra Club Bulletin* (now *Sierra*), and *The New York Times Magazine* for publishing other chapters.

Larry Ketchum, publisher and editor of *The Bisbee Observer*, for serializing the book, and for my column, "A Sense of Place."

William Eastlake and Marilyn Hill, for sending the manuscript and enthusiasm to Dream Garden Press.

Copyright © 1985 by Charles Jones

Manufactured in the United States of America
Book Design by Richard Firmage   First Edition

Library of Congress Cataloging-in-Publication Data

Jones, Charles, 1932-
   The gifting birds.

   Portions of this book were published in various periodicals.
   1. Life—Addresses, essays, lectures.   2. Space and time—
Addresses, essays, lectures.   3. Natural history— Miscellanea
—Addresses, essays, lectures.   4. Animals— Miscellanea—
Addresses, essays, lectures.   5. Jones, Charles, 1932-
I. Title.   —Addresses, essays, lectures.
BD431.J49   1985   113'.8   85-71151
ISBN 0-942688-27-9 (alk. paper)   Z39.48—1984. ∞

Dream Garden Press • P. O. Box 27076 • Salt Lake City, Utah 84127

CONTENTS

THE GIFTING BIRDS

I

THE GIFTING BIRDS

Places circle in my mind like gifting birds. They come down to perch on limbs and fences. They skitter in for hops and walks. They glide to rest in the still grass. No seeds in the garden interest them; they do not gyre down for any carrion. They leave gifts. Uncommon ones.

Quick perceptions I remember. Thrown lost on the desert. A sense of where my body was, moving on the earth, on the continent, on the beach, inside, moving away at the earthrise we call sunset. The sounds of a Mexican dance hall seeping into a kind of love all of us call young, but should not. A whimsical impulse to apply for a job as a waterfall keeper. The aching wish that a startled rabbit would not run. A sudden understanding of the grace and mischance of a pelican in my life.

Small, quick perceptions the gifting birds leave me. Morsels. There are so many—and I know them so well—that I collect them. Some I can put together, not as story or theory, but a sense of life, of place. There is the experience of time, that elusive movement of the universe which we measure, have a thousand definitions for, or say does not exist.

And in other parts of my collection, I can see how people make a place, how the soil in your fist tells you where you are. The way a water flea becomes a relative. Random perceptions become groups of experiences, pieces tumbling in. I have the feeling that I do not make them fit. They do. These are obsidian flakes from an arrowhead. Puzzle pieces without jigsaw cuts to define them. I merely placed them on a

9

shelf after the gifting birds left them and I picked them up. There was no message, no code, no mystical meaning. I sort and re-sort, and always it is the same. I know we have much to say. But I am dumb to the fit. The gifts are absurd, which is to say without sound, without reason.

The gifting birds were prophetic, though. They brought me places, perceptions, feelings, people, more often than I could collect them. Land like the Adirondacks was left at my feet. People like old Enos Ralston came into my time, or I to his. A dawn redwood, thought to have been extinct for twenty million years, became a life I met only once. Yet they were still silent. So I simply became a ruminant among those morsels on my shelf, among those gifts without occasion.

We are not born in a certain place by choice, but if we choose carefully, we know our places later. Usually, we do not do that. We drift. Near work, near home, an affordable place which turns out to be far too expensive for our sense of life inside. Our own eyes and ears, our senses, tell us, but only if we do not do what is demanded of us—or what we think is. For we do not learn anywhere how to pay that kind of attention, that careful, personal choosing of place. We take sensory, emotional, and esthetic notes all our lives, but nowhere do we learn how to live them, how to learn them, or even how to file them.

So it is a random collection. The bits my gifting birds left were pieces of places. A quick color, a friend in passing, a short luxury of growth. And some of these created in me a desire for eternal return to such places as Mexico, Hawaii and Arizona, to Vermont and Japan.

It was not until I found the place I live now that I could tell the gifting birds I knew them and had discovered their roost, where they went, and how they and I and everyone knows that all gifts are mutual friends of chance.

By stumbling on home, I learned how to travel. Place tells you, if you keen to it. A community is not just people, houses, and stores. An environment is not only mountains, rivers, plants, and the other animals. There is no good reason to separate history and natural history. A city is a wild place for a deer, natural for man.

These are parts, then, which come to the end of a journey where the gifting birds roost. As if a feral child had learned to say its life to us, to tie bone to word.

II

THE FINGERTIPS OF TIME

Plum Island Jack is—or was—something of a hermit dug in at a sand dune on an island near Boston. He may not be there anymore; he may not be anywhere anymore. That would still make me want to ask him about something he said. "Why, hell, there's no such thing as time." It took us about eight hours to get to the point at which neither of us knew anything more about time, and he said the same thing at least twice an hour. "There is no such thing as time."

Could be. We measure it, we say it will heal all things, we say it is relative, we say it is the great turning of the universe. Time fascinates everyone, metaphysicians to moviemakers, and especially kids without watches.

My first feel for time gone by, for the past, was when I was nine. Growing up in Texas, I had never seen anything manmade that was even close to 500 years old. What I saw on a trip to Mexico—the pyramids, the calendar stone, other artifacts—was a great mystery to me. Hundreds of years is a hard time to comprehend at any age. My mystery was that I could imagine people building things like that, down to a man cutting the stone. But I could not grasp his time away from me. Now, having learned quite a bit about life in times centuries and millenia before his, I still cannot grasp his time away from me. To this moment, it sends chills down my back—similar to ones that come with music and paintings—to hold in my hand a simple tool made by someone a thousand years ago. The chills are understandable.

What is more difficult is a sense of my own time, as a biological

11

being, as an animal here and now, at a given moment. The dance of time, going back and coming forward, is a step I can feel, even if the music seems muted. Yet for just one solitary moment to have no past and no future, to be just where I am, with no thought, is a discovery so simple that its proportions are huge. Our current state as human animals does not make us ready to know if that ever happens, that moment when we face no moment. As we are, always tied to something other, our own personal instant has a fearful face.

Perhaps the fingertips of time are frightening. To touch the past, to have it come forward and touch you, are perceptions all humans probably share but may not notice, may not assimilate in a way of life more given to short spans of time.

I think my sense of time was made wide and deep by making it shorter than it ever was, by the focus of a solitary place of barren heat and sand. That place, full of life, was as distant as yesterday, as vague as tomorrow. I did not really know that until later, on a fogbound coast with a heron's feather in my hand.

1. Artifacts and Amulets

In a house on the Pacific where I lived, there were times when I had to escape into the fog. It came so often and so thickly that it still lives buried in the books of my library. Manuscripts I sent off then were said to come from a sunken ship. In plain sight, the fog perched on shoe eyelets and turned them green, dove into walls and blackened them forever. The fog got to the fossae of the bones, and I could feel it inching up the vena cava to my heart. Perhaps to stop it, I dreamed of the fog. I saw gray at first, and then realized I was standing on the bank of the creek by the house. I was looking upstream into the fog. Out of it came a white goat on a raft. Goat and raft went by, a spot of white facing me in the white fog. I was delighted. Naturally there are almost endless ways of looking at that kind of dream. In the morning, I went to the stream to walk, thinking that the best way to make friends with dream and fog was to go out and meet them.

The goat was not there. As ever, in the summer, the fog was. It blessed me wetly as I walked along; the rafting goat left me. I went toward the ocean, looking at other things. Such changes in focus happen easily when we attend to our senses rather than a caravan of thoughts, or an untamed dream, which may best be answered by not asking a question. I merely attended elsewhere, foraging.

At a bend in the stream where a great beached log was, a lifeless bullhead lay at water's edge with no more effort left in it. The ones washed in here are small, six inches, with the outsized head of a sculpin or sweetwater catfish. Where the tides and the creek mingle, some life from both waters dies, taking clues too far, while other life turns back, following chance as well as migrant patterns millions of years in the untold record. For some on the sea side, a wave can be a

tumbril. From landward, a flooding stream fashions the last experience for a few. On the surface above the fish, angling coots know this
rhythm. They work the stream down and back, following, and hide
from the owls at night.

Along the creek, the path to the sea is lined part of the way with
berry bushes. The berries are small and tart, for the fog comes daily.
Even two miles inland, the fruit is larger, sweeter from the sun. Away
to the north of the path, ten acres of field lie in quondam use, leased by
the state to farmers now and then. It is prime agricultural land (which
is a classification, not a judgment), and once brimmed with artichokes. Agricultural interests hardly noticed that California's parks
and recreation department bought the land almost 20 years ago. For a
parking lot. Just to add to the state's loss of 20,000 acres of agricultural
land every year. The land is mostly dock and hemlock, a home for the
rare San Francisco garter snake after it lost its habitat farther north.
And the place is still good cover for the deer and herons and coyotes,
through and behind the seven-foot hemlock, made possible by the
giant agricultural apathy.

About a hundred yards from the beach on this day, tide met
stream with small, roiling walls of water and eddies on lunatic errands.
The swirls splayed out as if there were no known cause to them, no
moon pulling them, no place to go, some question about identity—salt
or fresh? Sometimes here, coming in with the tide, there are six-inch
bluebacks. And the adult steelhead, at the age of three or four, comes
back again to the same stream. It goes far inland to breed; it is often
two feet in length. But the steelhead, unlike the salmon, goes back to
sea. Local machos and matrons alike spear them at night or day, or
even collect them by hand, after the first heavy winter rains. Neither
game wardens nor steelheads favor it.

A little closer seaward, the tides go hard against the stream
grain, shoveling sand and whitening the water in a pebble tug-of-war
on the stream meander at the beach. At the apex of the winter rain
graphs, the creek goes straight and deep to the sea. At other times the
creek swings the whole beach, or lagoons in, can flood the entire sand,
or is barred from the ocean. It is then when, upstream from the bar,
the creek fills with algae in the summer months, and pinks and greens
float the stream in hydroponic mosaics. Fishermen curse, and I take

long walks alone on the path. On this day, the sedge looked dry in the marsh, with the height of the plant measuring the rains of winter. Only this time of year in summer and in the worst downpours in winter is solitude possible. The rains sweep across the place, gusting eastward, slacking west, and nothing stirs. "It is only water, friends," I want to say, but after all, I never worry about being blown off my wings or drowned in a mere rill, just about getting wet—which is only water.

Going back upstream, the fog flails around me but I can see through it that the sun is shining baldly on the Santa Cruz Range, or it looks that way. Usually I stop at the great log on the way back. The old pine dances a few inches in the highest waters, but may never move to sea. Across the stream from the log, a hill of striking symmetry is always reflected perfectly, rain or shine, in the creek at the log. I take that hill and fold it and refold it, unfold it in my mind, stare at it, leave it alone, glance at driftwood and rocks, listen and hope for the heron.

It was on that pilgrim stroll to the sea in a light glow of fog that I discovered the feather. Many times at the log, a great blue heron had flown at my approach. The slow bird would fly upstream and then turn back protesting that my admiration was unrequited. That day, I sat on the log, wrapped in the roiling fog as if waiting for a coven to join me.

Often, I could follow the heron in the air on its way downstream, but it was lost in grayness that day, and I looked down. There on the bank was a feather the heron had left. The bird was gone so quickly in the fog that I had a sense of loss. After all, the goat on the raft did not show either. But the loss was eased when I saw the feather there, and I realized I had wanted that feather for a long time, ever since I first saw the heron. I looked at the feather and thought about that. There was something in the artifact and the animal, something I could hold and see, close instead of far. Somehow, I knew the whole bird better, not so much just in small pieces of its flight. Trackers, I thought, do that all the time. The paw print, the scat, the tuft of hair speak to them about heft and direction, about the animal. Trackers can parse without losing the sentence.

It had been something like that with the coyote, except the spoor came after the sighting of the animal. My wife Molly had found a coyote skull on one of her walks in the field, and it was still sitting on her dressing table in the bedroom before its transfer to her studio. One night we were reading in that house far from anyone or any sound. The bedroom was warm and the light from our lamps cast out the window on a patch of graden. We were brought suddenly out of bed by a wailing sound, loud and close and fearful. A few seconds told us it was a coyote. We turned the lights off and inched toward the window. There by the garden fence sat a coyote, about medium size. Its head went up, the cry went out to a barely moonlit night. From a distance there was the yelping of more of them. This one did not yelp or yip. It howled long and clear songs. We watched it, just twenty feet from us, until it lay down, got up, and trotted away.

Though I am not a mystic, when the coyote left I walked over to the skull Molly had found, picked it up, and turned the light on. Why had that coyote come to our window? Why had it dared so close, sung there, and gone away? And why was I staring at the skull? Did I think I could read it, answer my questions from it? That night, I think I did.

Sitting on the log with the feather in my hand, I knew better. There are answers in the artifact, but I was not looking for those answers, not asking those questions. I was not asking any questions at all, yet. Both feather and skull taught me something more about the animal. Beyond what the artifact taught, it was evidence my imagination wanted to examine, a sign that perhaps there is such a thing as personal empiricism.

I wondered if, as with the feather and skull, we do that with the human animal too. Can we blur that strange line between history and natural history? I had started the day wondering about the goat and about dream time, then shifted to my own biological time, aiming my senses around, taking what they gave me without naming anything "wave" or "bullhead" or "log." Where do man's artifacts lead our sense of time and the earth? Images came in an instant from several years spent in Mexico.

A friend lived in San Juan Teotihuacan, where the Pyramids of the Sun and Moon are. Salvador was stationmaster there for the national railroads. The station was surrounded by farms, and every walk in the fields would turn up a *cabecita*, the little head of a clay figure. There were other fragments, too, of those amulets thrown on the fields to charm them into fertility (one guesses). The heads were a thousand and more years old. I remembered standing in the fields, mutely and with no thought staring at the first one I found. It showed the rounded edges of time, the face of centuries in the ground. Someone had made the amulet before Cortes came, someone had stood near where I stood and offered it to the crops. Pictures flashed in my mind of the man in the field. Scenes came of what were then tree-covered hills around the valley of the pyramids, which was a center of life, not a relic. The man—the animal to go with my artifact—was gone long ago, and I felt a sense of frustration at not being able to watch him in his time, as I was to watch the heron on its flight.

That same feeling followed me constantly in San Juan, especially with the new paintings uncovered in buildings near the pyramids. The brilliant blues and reds stuck to the surface like eternity, or at least the only one I could imagine. The artist's hand had been poised where mine was, his eye had been open to the colors of his work as mine had.

Still turning the heron feather in my hand, I suddenly thought of a shorter history, one made of memory as well as imagination, and much closer. My stepfather was a very great man. Though he turned no tides in his time, he was a man of thought and feeling and grace. We were friends in the way that old, very wise men are with young, very foolish men growing up. I got to know him for only three years, but he was a man you know forever. He has been dead for over 30 years, but I still have some of his things—books, pipes, photographs. Artifacts? Of course. The bond was made before they passed to me, but they are still signs of him, of my memory of one of the best of the animals called man.

I tucked the heron feather in my pocket and headed home. The bank rises on the way above the stream, and I stopped to look down. Always startled by it, I made out the doll at the bottom of the stream where someone threw it or a flood brought it. It will be gone one day, like the small concrete dam that used to be just above it. On the way up the bank, I veered on the hump of trail around the bees' nest and past the sweet smell beyond it. It is always there, the sweetness, like a sweet grass, but no one thing seems to bring it. I cannot find its source, the bearer of this mysterious pleasure.

It was a found day anyway. Feather and skull, *cabecita* and book, a white goat on a raft bringing me into the fog only to be missing. Both the goat unseen and the heron lost in the fog had left me a sense of artifact and animal different from knowing one by the other, goat by dream, heron by flight. That sense had moved into the fabric of imagination and memory, across the limits of categories like history and natural history, past and present. The heron and I were thrown here in this time, the goat in dream time, the artist in another one.

Walking by the creek, watching the lives going downstream and up, I usually felt moments of my own biological time. It is the sense of time without time, from which one awakens with only the present of the senses, an animal feel of being just exactly where you are and when. No other place, no other time. It leads nowhere, but is led by eye and ear, smell, the taste of air, the touch of wind. But that day, my own biological time *had* led somewhere. What I had wondered through—now, the dream, the past—did not seem to be thought or remembered or imagined. It was as if I had known them directly, just as I had seen the feather by the log, heard the wind in the hemlock.

There was no sense of being there rather than here, then rather than now. It was a strong feeling of connection, inward and outward, past and present, with no delimiting lines. It held. It crossed from the goat to the feather, skull, and heron to clay figures and a thousand-year-old painting and its artist to the treasures of a man I had known.

When I returned from the log and went inside the fogbound house, Molly was gone. She was at the general store getting provender and brandy to further make friends with the fog. I put the heron feather in her studio collection, remembering again to tell her about it. Then I drank some of the hot tea she had left for me. Later, I told her about the white goat and other artifacts and amulets. As much as I could.

2. Echoes Of A Singular Man

For a while, Enos Ralston was my holding dog. They came from the same time, he and the dog. Enos was born in 1859, when people still used a holding dog. It was a tool for anchoring a log while the adze or broad axe in the farmer's hand made a beam. Something to pin the beam, as log, to the ground while turning timber into lumber. A holdfast.

Ralston held me fast, part of me in his time, part of me in mine during the first year I lived in his house. The heron feather had led in many directions on one walk, but with Enos it wasn't like that. Many small encounters led in one direction, to Enos Ralston.

At first I only knew the house I moved into was 121 years old. Built as one room, it had five others added on, all on a step down, which preserves the roof line, but needs a hill or a large excavation. The hill was there, sloping down into trees over its brow. In the original room, the hole in the ceiling for the flue is still there. At the end of the room, the huge fieldstone fireplace overcame its heating inefficiency with the many incomparable, mesmeric warmings of every fireplace.

"Oh yeah," said Eric at the general store, "you're livin' in the house of Enos Ralston. Quite a man, he was. We had his tombstone stored here for fifteen years before he died. Didn't make much sense, but he wrote what he wanted. You know, he used to ride down here about every day for lunch at the old hotel. Ride all the way, five miles. He'd always bring something. Quail, vegetables, skunks."

"Skunks?"

"Well, he was a trapper, you know. Skunks, coyotes, coons, bobcats, foxes. He'd find baby skunks sometimes and take the stinkers

19

out. He'd bring a few down here and let 'em run around. Just like kittens. People would play with 'em."

When I told Gus, who had lived here 80 of his 90 years, that I lived where Enos Ralston had, he started to chuckle and shake his head. "Old Enos, he was a character." Gus was himself the best hunter in the area, even then at 90, but he said, "Enos Ralston was one hell of a hunter. He wouldn't talk, not a word, when he was out. Glare at you if you made a peep."

There was always something else to do, something else to write, but I would think about Enos once in a while, plan to find out more. From time to time, I would hear things. Ralston was a tall man, broad in shoulders, neither thin nor heavy. He was known by two characteristics. A pair of strong, gentle eyes. And red hair and beard. He wore, one oldtimer told me, "a little more than a goatee, like a doctor."

Ralston made most of his living trapping. He worked an area about ten miles by ten in the valleys and on the ridges near his home. Out on the lines, he would stay in deserted cabins or the barn on a farm. Wherever he slept, if he wasn't at home, he would pile up anything handy and throw out his bedroll in as high a place as he could get it. "Warmer that way," he said. The little dog that always went along with him made him even warmer. It must have been something special to him, that dog. It would trot beside the horse Enos was riding. When the dog got tired, up on the horse it would go—in a cloth bag with air holes, hanging from the saddle horn in bliss.

That was about all I had heard, here and there in conversations, before it started. I remember just how it did. Driving down for the mail one day, I suddenly thought of Enos Ralston. It was sunny, not warm. An ocean breeze mellowed it to a tepid fall day far enough before the "holiday season" so that I still felt good. I thought of Enos Ralston riding down to the hotel on his horse, in the sun. "Thought" of him is the right word. I did not "see" him among my imaginings. There was a certain feel, not as if I were Enos in my mind riding along unhurried to lunch. I felt as if I knew him, maybe was riding next to him, me in the car, Ralston mounted, talking to each other about the fineness of the day. It was just a sense, nothing definite.

The next time, only two hundred yards from the house where Enos lived, I stumbled on him again. There is today a summer place owned by a power company executive; there are two "cabins," both rustic palaces. On a walk one day with my wife, we started down the stream past the place, and as we waded along, I turned to look. There in a pond off the stream, floating in ferocious majesty, were two swans. Swans are a little formal for my taste, but what struck me was the pond itself. It was roughly a half-moon shape, or perhaps more like a teardrop awry. It had no business being there, except that the animal man likes to make ponds for swans. This pond was obviously older than the swans, though, and perfectly planned to sit by the house for the pleasure of its company. It was a graceful pond, and had some use at one time, its banks now in the luxury of grass. It seemed more serious than two swans idling.

Down at the store later, I told Eric about seeing it. "Sure," he said, "that was old Enos's hatchery. He raised trout there for years before it got flooded out." I smiled. Your pond is a good one, Enos, I thought.

Since the store is the only social center in town, that was where I heard most about Enos Ralston. Talking to a friend there one day, I was for some forgotten reason telling about the mud dauber, a wasp which takes gob after gob of mud and sticks it to any likely surface, such as under an eave, until a mud nest is made. If you take away each gob as the dauber puts it there, the wasp will continue placing new ones for a precise number of times. It will then put a dead but uneaten spider with dauber eggs in it right where the nest would have been—even if there is not one piece of mud left. As I spoke, an elderly man I didn't know came walking over from his barstool.

"That's just like a story I heard right here maybe fifty years ago. Couple of fellows used to hunt quail up on that ridge. Lots of 'em up there, but they flew in the berry patch every time. Maybe a thousand; made those two hunters damn mad. They went back to try again and again, but one time the farmer burned that berry patch. Well, these two guys went up there and flushed the birds and sure enough they flew for the berry patch, not knowing it was gone. Know what? The birds flew right down to where the patch used to be and sat there on the bare ground. Two guys got their limit, that's all, but they said the birds just sat there, like the patch was still there. Beats all. And I know it's true. I knew one of the hunters. No liar, not Enos Ralston. You know what he used to do? He was the best shot around, but he'd tell everybody his partner was better. I saw him do amazing things, like take tree branches off with a twenty-two. None better."

I looked up the ridge where the quail were and smiled inside. Good to meet you again, Enos.

A few more times I ran into Enos like that, such as with the rattles. I had seen them many times at the store, and they were giants for rattlesnakes in the area. Fourteen rattles and down. Of course Enos had brought them in. When he was riding down to lunch at the old hotel one day, the snakes had rattled at him from a ledge in a road cut, about eye level. It took him only one shot each, and that day he shared his lunch of rattler.

At some unmarked time, one I did not notice myself, I started asking about Ralston. Oddly, it turned out that very few people knew more than one story. So I stopped just meeting him by chance and went out to look for him before I had to move from his house. I was not ready for what I found of Enos Ralston.

Gus was the first one I asked, because he might know why people would tell "Enos stories" but not say much about him. "Well," Gus said, and there it was: "Some people think he was a little goofy."

"Why?"

"Oh... I don't know. I thought so myself, maybe, once. The time he hired my brother Jimmy. Went to help him on the lines. Jimmy got there the night before they were supposed to leave. Just after he turned in, Jimmy heard two blasts, Boom! Boom! and ran to see what it was. Enos was on the front porch in a nightshirt, holding a shotgun."

Gus raised the pitch of his voice to liken young Jimmy's. "'What happened, Mr. Ralston?' Well, old Enos answered him calm as you please. 'Nothin's wrong, Jimmy. Some nights I just come out here and let off two blasts. Tells the spirits I'm going to sleep now.' Next day, Jimmy told Enos he had a bum leg and couldn't go on the line. Jimmy limped away until he got over the rise, then he ran like hell for home." My front porch was a different place after that. I sat there and thought, Well, Enos, what next?

What impressed one man mightily was that Ralston told people around town that someday we would hear voices and see pictures from far away, right in our own houses. It is said that Enos read a lot, and maybe he just saw it coming. He also claimed that he could remember his birth, but no one could tell me where he was born. His parents came to San Gregorio from Winchester, Pennsylvania, but no one could remember where he was born. When I looked it up, the record was there. He was born in the house I was living in.

What people made of Enos in his day is hard to say, but it is certain that some of the things he did made the locals think of him as less than sound of mind. His stop at an inn nearby gives a hint of the way people thought of the old bachelor. To all who were listening at the bar, he told about riding a ridge at sunset one day. The sun and clouds set the sky to burning, as he watched in awe. Remembering the powerful, inward impression it made on him, he said, "In my mind, I was Christ just for one minute." The innkeeper grinned willfully and said, "Didn't hold your job long, did you?" Enos sat through the laughter, grinning slightly himself.

His neighbors laughed openly at him more than once. When he saw the creek blocked by downed trees and slash, Ralston would get a horse or two, some rope, and then go out to clear it, using a block and tackle, his power, and horse power. It didn't matter if it was his place or not. "Damn fool," people said, "he ought to be glad for a little dam."

Later in life, he apparently waged a campaign against flies, and everyone thought he had lost his marbles. He had a car in the late thirties, and would sit at the store on the fender of it. When a fly landed on his leg, he'd flick his knife down and chop it in half. Maybe he had lost his marbles; maybe he was hunting as much as he could in his seventies; maybe he was honing his senses.

He had done that before. Enos was obviously an eclectic in the meanders of his mind, and when he spoke of spirits or remembering his birth or becoming Christ for a moment, it was entered in the ledger on the side of his being daft. His sensory training supported that way of looking at him, at least in the minds of his age and place. Ralston set out to make his hearing more acute. He did it by concentration. Placing his pocket watch six feet away, he would spend sessions with it, clearing his mind and listening intently. Then he would move the watch farther away. He got so he could hear it three rooms away, he said. But he had to give up wearing the watch. In his pocket, the sound was deafening.

When he was 87, Enos Ralston moved to a friend's ranch because he needed help until he got over a kidney infection. But he never did. Five weeks later, on November 19, 1946, he died after sunset. The year surprised me, since no one remembered when he died or was born. But it should not have been a surprise. He was a man to live long, a man to hold on. Nothing really suggests that he was goofy. He may have been a seer or a mystic; no matter. He was clearly a man of his own. Somewhere, all of us understand or even remember that. Perhaps it is our holding dog, keeping us pegged to now and to then. He held me fast, dipping into his days, in a way I shall never quite understand. Perhaps it is too simple, that feeling that he was not a friend, but just that I knew him. That was all. Something I merely accept gladly without any spiritual or emotional explanation. He got me in his time and left me in mine.

The last meeting with Enos is something I can cherish not because I feel it or even understand it. It is the kind of thing that comes from long before his time and echoes into mine from ancient cultures. In his case, it tells me much more about Enos than it does of his time and place. Enos wrote it in 1931, or perhaps he quoted it, but he had it carved on his tombstone:

I AM THE PEER OF MY BEING
THERE BEING BUT ONE SUBSTANCE.
THAT MY MAKER IS ME
THAT MY MAKER AND I ARE ONE
THAT'S LIFE BLENDING WITH THE ETERNAL NOW.
I AM ALL THAT IS, THAT WAS AND THAT WILL BE. THAT I AM
WHO AM—OTHERWISE BEING COULD NOT BE.
ENOS RALSTON
BORN APRIL 27, 1859

3. The Lost Art of Solitude

The tailings of an abandoned mine pointed downhill to a sand road guarded with fences askew. Sun filled the landscape with a moving heat that had no brilliance. Low desert is faded in its highest heat, too much for the eyes, washed out in the sight. From a saddle between two small hills, I looked down on all that. Having been lost for two days, I expected nothing. But I knew the mine, and I knew the road. Yet I looked at it all with a dull eye. I got up, walked past the mine I had been down in before, cool in before, and then on to the road below. It was deep in dust, and I only stood there, confused, looking down at my feet. Quickly my gaze went east, along the road. I knew that road. Then I looked up at where I had been, back in the sere, lost country. Then I walked the road, east.

The sun on desert soil is worse than in furnaces. They are at least dark in the intensity of their crackling. Burning in the face, slits only in the eyes, I walked. Since I was found, since I knew where I was, since I knew where to go, since I had water and food, why was there no joy, no relief? I thought of doing cartwheels, but none came. I walked, puzzled.

Why was I not glad?

From where I met the road, I knew it was five miles to Sonoita Creek, and Patagonia, Arizona, was near there. Easy. I walked.

The first thing I found was that I had just learned the difference between loneliness and solitude, which meant I knew that solitude, unlike loneliness, is a matter of choice. I decided I was, among other things, a solitary, and was delighted to know it. There was nothing in my mind about a label, but only an awareness of enjoying company or solitude. It felt so simple.

25

Some years after that discovery, I was struggling through the sands on Plum Island, near Boston, alone with both the fine grains at my feet and the limpid heat overhead. Birds shared the day safely distant; I could hear them as they circled and fussed. Fine, dry sand is hard to walk in; one looks down in disbelief, up at the birds. I looked in between them, and, under a dune, sat a dark man by a dark door which led to a sand dugout with a driftwood roof. We looked at each other like two trespassers, poaching space not ours, and in the wrong season. We greeted.

His name was Jack, mine Charles, and we sat down. He and his house were dark from the sun and the burning of kerosene at close quarters. Neither of us were very friendly, and we enjoyed that, which made us talk more. Jack was a retired army sergeant from Brooklyn. I could not tell at first if he was a hermit in the sand dune or a recluse of the dunes, one being far and dug in, and the other near and simply hidden. Jack's stew bubbled in a fire on blackened sands. As he ate, he talked. The more he talked, the more puzzled I was. He was not really bitter about people, did not hate them, chide them, like them, seek them. One at a time was fine. He simply could not stand more than that, and preferred none, no people at all. "I hate it every time I have to go to town to get my check and some food. Hate it."

"Why?"

"My life is mine, that's all." Hermit or recluse, Jack was no closet loner, and no solitary, either. He had no choice.

Even later by a few years, a friend came to visit from New York right into the California woods where I lived. Surely it was the house of a solitary, alone and enclosed in the redwoods—but near a fine city. The first day of her visit, I had to go to work and leave her there in the woods. When I came home, she came running across the long porch under the trees. She leaped down the steps and shouted, "I'm so glad you're here!" She yelled, "This is awful!" What was? All alone. All day. Two miles from the nearest person. She had been alone all day with only the sound of the wind in the trees and jays rasping—along with a radio. But she had come from a city of millions to a house of one, and she was terrified. The barely heard hum of a city's background, the rumble of a passing bus, a corner store, a city smell, anything from her home would have calmed her. After two weeks, there was little difference. When she got back to New York, she was comfortable again where she was never alone; but she was steeped in loneliness she could not understand. Plum Island Jack was, as one says, lonely in a crowd, my friend crowded when alone. She had no choice.

What of those who have a choice? To give the solitary any other quality than choosing to be alone or with people would weigh him down with definition and category, which he cannot have. He cannot belong to Solitary Liberation or go to Solitary Centers. The art of solitude is only individual, and has no particular place.

As a veteran solitary, it might be expected that I provide some notes toward the art and its practice. I might begin, "Take an empty place and a full person," but that only sets a stage, and pretentious prescriptions are no use. It is one of those rare pleasures of the solitary in these public days that his art can only be his own. There is no teaching, only learning. That is a richness in our age, when solitude is a lost art. It is clearly related to the contempt our society holds for people as they come—one at a time. Only in groups are causes recognized, and there is nothing called Human Liberation, which would no doubt be as glib and groupy as all the rest.

Which means that solitude is not only a lost art, but is suspect in our time. The shocking fact is that the solitary chooses his interludes of aloneness. Is he rejecting people or society; is he sinister; a threat; immoral? Of course that is absurd, or silly, or both; but all of us know someone who thinks that way. One would never say, when asked at a party what one does, "I am a solitary," or "My interest? Yes, it is solitude." The delights of solitude are rarely spoken of, and places of marked solitude even more rarely understood. Man the meddler cannot leave solitary places alone. The wariness with which we approach solitude is clear when we think of attitudes toward people like backpackers and birdwatchers, who are seen as quaint oddities or nuts—and sometimes are, of course.

But here are human animals willing and even anxious to go out into remote places just to be alone in the unpeopled splendor of the earth, their home. Where? Not in very many places, even though solitaries practice their art in every place. The Wilderness Act of 1964 has all but fizzled out in favor of urban parks, no parks, or scouring out the solitude in the name of profit. And solitaries are a minority. We do not nurture solitaries. In our schools, grouped around our TV sets, and computerized to the very brain stem, there is no time, no room, no button to push, no connection, no allowance for the solitary.

In our cities and suburbs, solitude is rare enough. Lone people are suspect, watched for quirks and improprieties—and especially crimes. But even where there is a natural opportunity for enjoying being alone, there is no let-up. Once I had occasion to write to all the national forests, parks, monuments, recreation areas, wildlife refuges, and state parks about wilderness areas. With the exception of fewer than a dozen replies, all were cool or hostile about backpackers, especially in small numbers. The reason given was that backpackers, birdwatchers, wilderness fisherpersons, and other solitaries were simply too small in number and did not deserve a place to go. In the same replies, almost every answer said the wilderness areas were overused. Ordinarily, that means expansion of facilities and services. In this case it meant not adding more wilderness, because it serves too small a group of people, but adding more rangers to control the crowd. Some wilderness area rangers were uncomfortable with the whole idea of "loners" too deep in a wilderness area to supervise.

Closer to home for me, there is a 5,700-acre park which the Parks and Recreation Department of San Mateo County, California, thinks belongs to it. In this undeveloped park, these bureaucrats wanted to put an extravaganza with shuttle buses, parking lots, a four-lane access highway, and circuses. One of the circuses is a redwood tower, which people can climb to see above the redwoods, and disturb all the nesting birds, including the great blue heron. Left alone, the park could be a perfect wilderness area, even larger than the minimum study area of 5,000 acres in the Wilderness Act for federal wildernesses. The cost and environmental damage would be very low. Yet the environmental impact report for the park dismisses such a possibility. It would serve "too small a spectrum of users." Whether this means there would not be enough people to crowd the park, which is patently absurd, or that backpackers are all too much alike, is not clear.

Just a few miles from where I live now, the State of California was also ripping out a solitary place to insert a circus. Thirty acres of prime agricultural land were bought in 1965 for a parking lot, campsites, ranger houses, an information kiosk, trails in the brush (there are no trees at all), camper hookups, and picnic tables. None of that is related to the beach it was supposed to serve and not even very close to it. The creek trail to the beach used to be a fine one for the lone hiker, but no more. Recreation does not recapitulate re-creation. Only lack of money has stopped these developments.

And solitude? That was long ago and in another country. Indeed, one often speaks of "entering" a wilderness as if going into another country, and not without reason. Few people know what to do with themselves in that foreign land of solitude. So it must be tamed, trimmed, neatened up until it looks familiar—crowded, cluttered, littered, riven with the sound of motors and radios. Then there is no fear of solitude.

But solitaries can live anywhere, city or country, be anytime, now or then. Since I can instruct no one in the art of solitude or the personality of the solitary, I can only testify to my own discoveries, made on the simple level of an experience between the land and one human being. Such things as sunsets taken alone can expose us to solitude. Pain can, too, for it is always and everywhere done alone. My best lesson, though not the first or final, goes back to the heat of that dusty road in Arizona.

In the dark of moonless nights on the desert, it is easy to get disoriented, lost. I was trying to drive to Amado, Arizona, on a dirt road, which began—or ended—near Patagonia. In those days, the early fifties, you could never be sure there would be a road there at all from time to time. My idea was to drive the route first, west to Amado, then hike back east along it in a line a little north of the road. I started out in late daylight on part of the road I knew. The earth turned from the sun as I headed into unknown land. At one point, as I learned later, the road branched gently north, with the Amado road continuing west. In the night, I went with the north road. Mystified at not finding Amado, I finally stopped to sleep.

The next day, I decided to start east without Amado, and did, and was too far north of the road I had come on. So I started out lost. It was marvelous: as tough, beautiful, and gentle as you could want, sered brown under blue. Of course I knew where the road was, water at various places. Just over the hill a bit from where I walked, just a little south. Ten miles, it turned out.

The first time I looked for the road and couldn't find it, I knew what had happened. There are hills there, ones you see from the road down below. The ones I saw looked like mirrors of those until you walked up to the glass and saw all the reflections, one going into the other, all real. From where I was, that was no help. I walked. I stopped. Where was I? Where was the road? South. South, of course. South I went. More properly, I went southerly, tending east, for that was my destination, if I was where I thought I was. Going straight south to the road just to find it would have taken away this bonus of land I had never seen before. I was farther away from the road than I had planned on, but I had food and water. On the second day, I dipped south again for a short stretch, looking for the road. It was not there.

All my life I had scoffed at getting lost, but even before this trip, the Superstition Mountains had taught me not to laugh. Until you are there, you don't realize how ten miles north or south or any distance in any direction can look like any other. Weavers Needle, a landmark there, anchors you in space. Then suddenly you can't see it, but you see a familiar rock, a known canyon. Just as suddenly, both become mysteries, out of place. You begin to wonder about the sun itself. Where, what time, what direction? I remembered that, and finally getting out. The thing was to feel the place, to get out by looking and knowing what your sense and senses told you. No problem yet.

So I walked. Southerly toward the east. The desert there is not sanded dunes. It is grit and rock and stone, and it grows ocotillo, saguaro, and cholla cactus, lizards, birds, coyotes. As well as an occasional dram of terror; but I knew that and it was in the safe part of my mind while I hoofed up and down hills, ridges, washes. Nothing high, only scrabble and slide. I stopped to look through the hand lens at likely rock specimens.

On the third day it occurred to me that I might be lost, as in not knowing where to go. I already knew I didn't know where I was. Where else to go, other than where I was going, for the road? Stop and look. It seemed right. No fear yet, for the desert was still friend enough. That night I looked at the stars a lot, decided it would be a mistake to worry about getting back on time, and slept heavily.

The next day was different. I had no sense of time. No, not so. I had no sense of past or future, and the present was only what happened under my feet, squinted into my eyes. Neither time nor anything else passed. I was not connected, not with time, not with another person. I was right there where I was. I walked southerly toward the east. At my pre-noon hideout under a cholla, this is how I felt: Thrown. I was splattered flat on the desert, impacted in the sand by the sun. I drowsed in and out about that, vised between the heat of air and earth. Then I began to wake up.

The first thing was a great feeling of comfort, with no sense of danger or loneliness. Suddenly, I felt a sense of excitement. As it is with a sight, a fresh vision, I wanted to linger, stay, not give it up. I walked around the rest of the day, exploring, seeing things, hearing, connected only to what was before me. I was on my own biological time.

Then, at that time, I was not ready for the feather left by the heron on a later stream; no spoor could have led me anywhere, not to place or time. I would not have seen an Enos Ralston, holding me in two times. The gifting birds had nothing to leave but the momentary smell of the dust at my feet, heat on my head. Sensations without time gone or time to come. It was just between my place and me, with great affection. That night, looking at the stars, I laughed outrageously loud.

There was no danger. A little water left; plenty of food. The next day, just wandering up through a saddle, I saw the mine with its tailings pointed to the road. When I was down there, I stared east along the road longer than I needed to. Only five miles to go, and I did not want to go them. It grew slowly on me, walking, and came greatly later at Sonoita Creek. There, I lost my shyness at being found, wallowed gleefully in water, mud, and my still-secret survival. That night, friends celebrated the trip with me, and we were all glad.

Yet I knew why the sight of the mine and the road had given me no quick joy. It had wrenched me too quickly from the finest solitude I had ever known.

That day on the desert, thrown on the sand in magnificent solitude, came in its own force to me only later. An early gift from my private birds came to me once when I was reading *Hamlet*. "To be or not to be" is either so familiar it has become trite or its words are so etched on our brains that it has no more meaning. It occurred to me suddenly that there is a kind of corruption about that soliloquy. Esthetically, we savor it. Yet I cannot already be, cannot be alive, and ask that question. I would have to say, To continue being or not to continue being—which does not scan. But "continue." It tugs at me from the future, where I cannot live. It reminds me that I have been, in a past which is gone. When I thought about that, I knew that the moment on the desert gave me an intensity of a present I could not ask questions about. I must think in order to ask a question. That instant let me touch the fingertips of time.

Without that moment, the heron's feather might have been just a spoor. But the markings of time trace further. That time on the desert when there was no time was as short and as long as seeing the man who made the *cabecita*, as real as Enos Ralston taking me out of history.

Enos came into my time. I imagined him around the place, just now and then. Usually, I think I would have given him a character beyond what I found from those who knew him, beyond his track across the records. But I was not going into his time; he was coming into mine. It was neither the substance nor the spirit of Enos Ralston. It was the signs of him, and the memories of other people. It was not a silly immortality, but the clear fineness of a mortal. There is no longer an Enos Ralston, but I still like him. I did not know him then, but I like him now. Perhaps it is understandable as the opposite of the first woman I ever loved. I do not love her now, but I still love her then.

Which may mean that there is no way to tell time, to name or define that invention we have posited and used. I did not care about it or know about it that day on the desert. Some years later, while agonizing about things like free will and determinism as a philosophy student, I wrinkled my brow and brain about time, too. I thought the fine filagree of time's ideas was marvelous fun (which one is not supposed to think), and very serious.

Later, when I had put these and other brief perceptions together, when they had become full senses in this place where I live, the gifting birds would tell me that the fingertips of time touch lightly and etch with a fine point.

III

A CERTAIN MOVEMENT, A CERTAIN STILLNESS

We notice those perceptions of time—the past, our own biological present—only if we pay attention to them. To mark the movements of our universe, of our own bodies in space, also takes attention to both smallness and largeness, a perception of reference, of the relative places in which we move.

But "pay attention"? Attending is more than simply noticing. Trigant Burrow, a remarkable but almost unnoticed American psychiatrist who worked in the first half of this century, explored the notion of attention. Burrow wanted to map, as it were, not the mind of man but the mood of man. To him, the healthy person was one who had the natural, primary relation to his environment still intact. Burrow considered attention as the concern, the care, of the organism as a whole related to the environment as a whole.

In Burrow's terms, successful attention was "cotention," a condition in which the organism achieves a tensional balance of cohesion between itself and the environment, the immediate and mediate world around it. It is the feeling we have when we know we are exactly where we are. Everything "fits." Our body and the fabric of the world at our fingertips work together. When that fit is distracted, discontinuous, when there is a dissociation from our own biology, Burrows says we are in a condition of ditention.

Perhaps most of us live in a state of ditention. We do not notice our own biology, our animality, each of us, one by one. We do not

34

sense—we do not attend to—our own perceptions. We see many of the possibilities we have only after a quick choice tells us to look back. At a superficial, even casual level, we choose mates and places and jobs without having discovered all the possibilities to choose from.

Here at the small, quick, one-time personal perception, we begin to miss. Time is not lived but counted. The mere movement of our blood and of the stars somehow escapes us. These perceptions of time and of our movement are so minute as to be missed, ignored, elided, thought minor. And so they are intensely valuable, to be hungrily attended.

The gifting birds have left me a few cotentive pieces of movement. The immense and total movement of me as a human animal and of my palpable world are left to me on a beach. Paradoxically, living space becomes clear from the sense of unlimited space on Fujiyama. And on the shore of the Arctic Sea, the gifting birds define where I am by showing me I am in no place at all—and everywhere I could be in movement and in stillness.

4.   A Continental Passenger

Usually when I go to San Gregorio Beach, I have a mood and must investigate first to see if the beach matches it. These favored sands are only a few miles away, so I go to see; there is no other way to tell. If the beach is well peopled and I want to watch visitors on errands of entertainment, I drive in. At times there is no one there, and good times await the solitary, who, unlike the recluse or the hermit, chooses time alone. If it is glaring sun I want, my wife, a fan of sunlight, must come with me. We dig sand worms and she casts for fish in the surf while I do some professional brooding, lightened by shimmering sand and sun. I lose myself and sometimes don't know if I am sitting in the sky watching the sand, or the other way around.

Often, I go just for the fog, which is plentiful many mornings and afternoons of the year. I walk along the beach, with the gulls scattering as they judge my trajectory coming near. The unkempt fog spirals up, washes in and out. Everything is movement—swirl and flight and step after step, with the waves awash, shouting or whispering. As the fog lifts, the gulls fly, and that foggy deception of the senses disappears above. I sit on the beach, stone-still myself, and watch the sea come and go.

One day on the beach I did not sit like a stone. I sat like a passenger. That morning at the ocean, the surf was stillborn. It never got a chance. Flat as a glacial lake, the Pacific just lolled around and back into itself. When it is like that, I think of Antoine Roquentin in Sartre's *Nausea*. Under calm seas, Roquentin did not see tranquility. What of those teeming fish, that eternal struggle, the silent falling of the dead? To imagine it makes the mind spin with uneasy mystery and aching curiosity. To think of empirical knowledge embracing it all is such a distress that a picture must be created. As with a small

particle discovered by a physicist, poetry, a vision, takes over for the
next, even smaller particle until it too is verified—or not. That lonely
vision is enough for a certain vertigo of the mind.

Roquentin tells me about nothing but himself; fine, but not
enough. If I turned to Tethys, Greek mythology's mother of the
oceans, I would have to be skeptical listening to a mother tell me of
her children. No, I wanted a picture of my own, a sense of feeling
where I was and where everything on the ocean's floor was, where we
met, and how we moved right there on San Gregorio Beach. My
picture settled down to bedrock, the sea floor, under the benthos.
Surrounded by the movement of tides, gulls, and fog, I saw first the
steady land under the sea. But suddenly, I had the picture I wanted to
unseat me from the beach. With all the movement around me and in
the sea, I remembered I was also sitting on a moving continent.

It is a recent idea in man's history, this thing of the ocean floors
and continents moving. The thought of the earth moving around the
sun was difficult enough. It does not look that way. An effort must be
made to feel the earth turning toward or away from the sun at dawn
or dusk. The sun does not set or rise. We turn to meet it, ride away to
leave it.

The ocean floor and the continents also move. Not only does it
never look that way, it never feels that way. If anything is solid, it is
the earth. That does not account for the Finnish coast on the north
Baltic Sea rising one meter every century, the north shore of Lake
Superior rising two meters every century. The earth, like the soil and
life on it, is not only turning, but moving everywhere.

Francis Bacon was the first to mention that a map of the
continents looked like pieces of a jigsaw puzzle. That was in 1620; not
much happened to the idea. In 1858, Antonio Snider-Pelligrini suggested that the continents had moved. Their motive force in his
scheme was The Great Flood.

It seems at first unlikely that Alfred Wegener would present a
coherent theory of continental drift. He was an astronomer, a meteor-
ologist, a balloonist, an explorer, but not a geologist. Yet many ideas
make their way, usually slowly, from outsiders into sanctity. In 1915,
Wegener published *The Origin of Continents and Oceans*. For almost
50 years, his work was mostly either ignored or ridiculed. In the 1915

book, Wegener saw the continents as originally one land mass, which he called "Pangaea," from the Greek for "all land." Wegener cited geological and fossil evidence linking formations in South America and Africa. And he relied on the idea that the earth is a thick fluid, which, like pitch, breaks under a sudden force, but moves slowly under such forces as gravity. That idea did not work out, but for Wegener, the continental movement was caused by the rotation of the earth, acting as a fluid would.

Before Wegener's day, horizontal movement was ruled out for the continents. Vertical motions—the sinking of land bridges, the formations of horsts and grabens (upthrusts and downthrusts)— were accounted for by gravity. But the substratum was solid. Wegener proposed that the continents migrate horizontally through the substratum, a viscous fluid, not a solid layer.

At first, researchers who were excited by Wegener's ideas began to write a poetry beyond and slightly different from his. Then the evidence trickled in, beginning after World War II. There were increasingly sophisticated probes of the sea, and empiricism started to nudge poetry. The great rifts of the oceans were discovered to be the source of igneous materials welling up through the long fissures, pushing out along the sea floor. Core samples and magnetic readings confirmed the idea of continental drift. The deposits farthest away from the rifts are older than those near them. Contrary to what was believed before, the oceans are younger than the continents. Magnetic changes fit the same timetables. The earth's magnetic field goes through reversals not yet understood. These can be traced, though, in different ways. Even ancient pottery which has been fired shows an orientation of particles in the pottery with relationship to the earth and its magnetism.

The current theory of the motive force of continental drift is that the upwelling magma is moved through the rifts by convection currents inside the earth (a theory with still some ragged edges), and that continents move on plates defined by the oceans' rifts. The intruding flows on the sea floor are slowly pushing out from the rifts; slowly moving mountains, creating earthquakes. Slowly. It has been estimated that the voyage of Columbus today would be longer by a football field, the Atlantic Rift having pushed the continents that much farther apart. The rifts of the Pacific urge the sea floor up and inland along our beaches. Here at San Gregorio, the push is through the valleys against the Santa Cruz Range, shoving the central valleys, and raising the Sierra Nevada slowly higher.

Odd, this thing of change and movement in our world. We want the Rock of Gibraltar, something solid, Rock of Ages. In a short space of time, to watch a night-blooming cereus open in minutes is one thing for a flower. To feel that the very ocean floors and continents are in motion is another. It is as undetectable to most of us as the cereus is observable, but we think of flowers as fleeting, the earth abiding.

As a geology student for a time in the 1950s, I was in a school dedicated to hardrock mining, and I never heard of continental drift. But along the way, the most exciting things I learned were about the face of movement and of change—faults, synclines and anticlines, sedimentation. Most of them were ancient, local movements from time past, or would move again in time to come. But the present, time now to my senses, was only told by a geology pick clicking away at a frozen moment. Continental drift is a different story. Even at this very moment, the continental plates are on the move.

My excitement had come not just from poetry or myths handed me, not only from Tethys or Roquentin, and not merely from the work and lives of many geologists after Wegener. A scientist would scoff at the word, a religious fundamentalist would think it blasphemy, but the idea, the feel, of drift was a revelation I wanted from the earth. I knew the earth moved about its axis, around the sun; that the core was seen as molten. Now, with drift, the dead crust came into the fold of movement.

That day on the beach, though, told me about more than an idea. And it was different from imagining a darting perch in the sea, watching sand blow, just thinking of a creeping continent, each a thing apart. All at once, in one small perception, I could feel a physical continuity. I could see where I was, sense my body's migrations and those of everything around me—the gulls above, the grains of sand at my fingertips, down the continental shelf into the deeps. It was one single moment, a simple, animal perception of where I was and when, moving among movements, changing among changes.

5. Fujiyama And A Space Of Difference

In some brief and innocent days, Japan was for us folded into the mysterious East. That was not so, of course. Every place has its mystique, its own elán, or several, woven indelibly with commonplaces in hiding. One does not see them, or they are hidden in curious ways. They are the everyday furniture of the universe, but extraordinary to the visitor from another universe. The smallest nuance blossoms with a certain dazzle not apparent to the casual user, the daily viewer.

So I sat staring at the bowls with some of the mirth of wonder. The wooden bowls were stacked in holders along the tables under a shed at the base of Mount Fujiyama. It was early evening, and scores of climbers were having rice, fish, radish, seaweed, soup, before beginning the all-night climb to reach the top by sunrise. I was the only one there who was not Japanese. Some students near me offered words, offered food, and we traded. They knew—and somehow liked— my exotic C-Rations. I was to climb with the students to the top. They were my teachers.

There was never a time in Japan when learning stopped. That country which not long ago seemed impenetrable gave up its lessons while guarding its mysteries, which is true of every place on earth. An alien pace, a pulse rarely taken, is everywhere obscure. But attending to what happens on doorsteps and in stores, as well as in temples and museums, makes everything enter the weave. There were ways, though, in which that patterning was hard for me to find in Japan.

In those times, I suppose I was technically a soldier of the occupation, though the thought and fact never came up. My own reason to be there was for Japan, to hear it, see it, smell it. But there were barriers between Japan and me. Enlisted men could not wear civilian clothes, so the uniform was always there. It would have mattered little. A tall American kid is just that. I was certainly obvious, and a curiosity, but I never experienced any overt hostility. Such acceptance—or perhaps it was practiced control—creates an atmosphere in which one is not timid to learn. It happens in a conscious way, and you find your senses unwittingly attending more.

It was first the space of structures which impinged. More like an aroma, just a waft at first and then a pointing of the nose, homing in. The attention begins to work. Obviously, I was too tall for doors. Walking around in the Imperial Hotel, the Frank Lloyd Wright version, I had to be careful to duck. In other places, sitting on the floor either properly or improperly at the table, I felt like six feet of legs and two inches of body.

The subtleties of difference came more slowly. Before Japan, I had been quite a bit of time in Mexico, and some in Canada. Any time in another country tugs us suddenly into a powerful awareness: the experience of difference. We become no longer the habitués of our daily perceptions. There is no more "just like home," which becomes a small piece of impossible and ignorant desire. There are those who are uncomfortable with the experience of difference, those who even suffer under it. And some are nurtured by it, sustained by the experience of difference. That is what Mexico, especially, had done for me—given me a taste for the broth of difference.

In Japan, though, a new measurement was to be taken. Language, people, food were different from anything I had ever experienced. One expects that. What I did not expect was the difference in space.

In a country of so little room, space was created as a series of refined, distilled esthetic perceptions. On purpose. Space is not an emptiness to be filled by accident. Space is a positive, planned dimension of perspective and of attitude, introducing a point of view. Cities have many houses and little land. Look out in the Japanese garden. At the end, there may be an arrangement of stones with small trees, the bonsai, near it, and larger growth increasing toward you. It is the mountain far away, with the distant trees small. Outside and inside

have been sharply separated in a seriously playful deception, a vastly
honorable ruse. Except that it is all there, all "real." The owner of the
garden has pleased and deceived his own eye for an elegant moment
of unknown duration.

The niche, the small room with a bareness to one's eye did not
make the room larger. The space was larger. Yet it was not an
expansive sense, but rather a serenity of proportion, a perception
created by minds aware of all the senses. As a tall man, I did not fit the
spaces. As a human being, anyone would.

It would be perfect if I could now say that what I was doing there
in Japan at the army intelligence school was an enlightening contrast
to my discovery of a space of difference. But I had not got that far.

At the school, I was learning how to look at aerial, stereo
photographs and find tiny things in them. Even from 40,000 feet, it is
possible to spot a person walking along a road. It is far more difficult
to look down a garden 25 feet and see a tree miles away. Perhaps it is
like the very old joke about walking into Abercrombie and Fitch for
outfitting for an elephant hunt. One is handed a pair of binoculars, a
jar with a lid, and a pair of tweezers. Once in Africa, and the elephant
is spotted, one merely turns the binoculars around, opens the jar, and
puts the elephant in with the tweezers. Thus a play on reality in the
Western world. If your eye can see, though, in Japan the elephant
already fits the jar.

There is no moment at which that awareness happens. It washes
over you in gentle, barely conscious tides of days. It is a sense of
difference which allows you to dismantle, slowly, your own cere-
monials. And then chance ambles by your attention.

In Asaka, the town near the intelligence school at Camp Drake,
there was a bar and coffee house. It was quiet and not like the usual
swilling places for the troops. The walls were lined with record
jackets, music of all kinds. You could look around, ask a waitress to
come with you, and point to things you liked. In a short or long while,
you might hear all or part of it. There were not very many people, and
I was curious about classical music, so I learned a lot, just choosing at
random from Vivaldi and Mozart and Brahms. It did not seem an odd
way to learn; if I had known more, it would have. Evenings there were
very special, not only for what I learned, but for the fact that I was very

shy at the time and did not feel comfortable at the GI places. There came times when I had to travel because I heard of good places from the people there. I went to Lake Chuzenji, to Nara. And then a waitress at Asaka told me how to get to Fujiyama.

It had never occurred to me that thousands of people climb Fujiyama every summer. Without being aware of it, I may have thought that mountain too sacred to climb, though I had done considerable climbing. The photographs of Fujiyama show its perfection in human, esthetic ways of seeing. Perhaps it was too pure to climb, and it is not, in any case, a climber's mountain. It is a long, very steep hike. It is an act of such deep simplicity that one may not notice until later what has happened.

On the way there, the train was as always—then—full of friendly, curious people. They looked at me obliquely and smiled at me directly. To Western minds there is something ominous about that, something lurking behind either facade. Of course there is, all over the earth. There is nothing extraordinary about bad temper, ill will, or evil hiding between us. But the Japanese, then, were at least pleasant to one another and to aliens such as I, met in passing. In other places, the mere act of passing is written off, hostile or ignored, indifferent. Naturally there was the element of looking so strange. As train mates, the passengers and I passed the trip with a pace of gestures, sudden conversations ill-understood, and the silence of shyness.

I began to wonder about the climb. It was hardly like just another mountain to go to Fujiyama. On the way to a sacred place, I was an unknown and unknowing man among pilgrims who understood themselves and where they were bound. Fuji-san is an isolated mountain, and has the symmetry of an existential geometrist's dream, a proportion to define cone volcanos. I found it utterly impossible to detrain and ask where the mountain was.

There was no need. Down off the train, I saw dozens of others with rucksacks and what seemed might be climbing clothes. There is no special equipment and there are no requirements needed to climb Fujiyama in the summer.

I followed the crowd, as one does when unable to ask where a sacred mountain is. And we all joined together in the large shed with no walls. There were hibachis going, and those stacked, hot-food holders came out. Gladly, I passed my canned and packaged food around, then tasted things I could not have named but could have eaten forever. It was only here I learned how one climbs Fujiyama in summer. After the food is over, there is no need to look for a place to sleep. You begin climbing at about 9:30. Many people had started. Up the slopes of the cone went light upon light, waving past sight. Each climber judged the time for sunrise. At the top, there would be ten wood-burned symbols on the hiking staff you got at the base.

Through the night, on the way, there are ten stations where you can rest, eat, drink. At each station you get a brand on the hiking staff. The staff is useful on the slope as a balancer, so that you neither leave the mountain nor hug it too hard. It is a comforter. Then, with the wood-burned brands of each station, the staff itself becomes a symbol. A fixed memory. A pledge done. A place known, if in darkness and with a dedication that might not be speakable.

We began. My teachers, the students, at first kept wary track of me. I think it occurred to them that I might fall off the mountain, and one can do that. A scramble in the flicker of lights is bad enough. In the steepness and darkness, I struggled against an ill-met fantasy. I thought I ought to be dizzy, but something—ear, the eye subdued—said I was not. At times I wanted to lean out, away from the slope, or I wanted to fall in and grab it. There were no handles anywhere. And I already knew better. The balance, the body in space, knows more than to think.

The trail was deep in the notch of ridge where we traveled. It was humped and swaybacked, eternally steep. Words I could not understand entered and left the place where I was, swirling with the crunch of rock and clunk of hiking staffs. The staff was an orienter, something to touch the earth with when it warped and tilted and moved. At each station there was a small hut at least, and the brands in the hibachis to care for the track of the climb, to signify it. There was, at least as I could sense, no question of "proving" the passage by the brand of each station. It seemed to be a matter of the esthetic completion of the staff itself. Such a thing may sound silly now, even in Japan, but the staff places you in space as you climb, and then it becomes a counter, a teller, of where you were. It is at the same time an object personally and esthetically venerated but with a meaning

inexplicable even to climbers who have several. Of course this climbing of the great cone and carting the great staff have droll psycho-
logical explanations. Such are the possibilities to commit theory, or to
giggle. What is really there is man the animal in the fullness of the
space given.

It is true on the mountain. All along the way, people of every age
go before, come behind. No one competes or pretends to manhood or
womanhood. One steps off the trail to rest a bit where there is room, if
it is only to stack one by one by the trail. The faces flicker as they pass
in risible threes and fours or solemn, silent lines. And at any gap you
want, you file back in line. Again you get your mountain legs, signals
from the inner ear, nothing but confusion from the lambent torches.
It is the steepness, all night long. As it is at sea, there is no place to go
but on. Up, here. Eyes on the ground, standing in, standing out,
stopping for a rest or a brand.

We were not all the way to the top, still moving single file, when
a hint of light rolled in from the east. Everyone stopped and turned,
doused lights, and stood waiting in the quiet. First the sun, out beyond
anything, far. So far, so starkly alone. Then the clouds; the light sank
them even farther below us. Down there, out and around, nothing
rose above. Only sun, only clouds, only light. We were people alone
with little land on that impossible cone with no trees.

A diffuse light came first, silver at moments. Then the orange.
In seconds it was too much fire to see. The clouds turned to humped
waves of shadows and lights, tilting at pure space. As for the way we
see: The sky was higher than it could be, enormous, over, above,
around the sun. And down, the clouds, fixed like sea, went lower than
anyone has seen neap tides. I stood there with it on the pitch of the
slope. Drooped from the climb, I knew exactly where I was.

And then I sat. Without time. My friends were gone, and people
filed past. And I remember that I was hiking again, the private
moment gone, and the light after so much dark gave space a remembered look. After almost eight hours of not knowing where I was, had
been, or was going, I felt a special fondness for the slope in light. The
special concert of eye and ear for telling you where you are was
working again. And I could now see Fujiyama so closely that I could
not see it.

The topmost station is larger than all the others. It seems to have space inside just like flatlands. There were sliding doors, lanterns, hibachis, voices, food, sake, beer. Much talk, with the blessings and lilts glad with what they had done. Even there it fit, and there was comfort, and there was no one to say that what is serious is separate from what is pleasant.

All of us lingered over food until it was time to go to the crater. I had found the students again, and we went. It was impossible to tell how far the crater rim was across, or how deep it was. This mountain does not divulge references. With no trees, with just patches of snow and ochre earth, grey streaks, there are only the choices of far and near. The crater appears to be either 150 or 1500 feet deep; it is 732 feet down. The rim is easier to judge, though it seems smaller than it is, with a diameter of almost half a mile. I wanted to go around it, to see and to walk on level ground again, atop a mountain of 12,395 feet. But I had a train to catch down below. A gross absurdity.

My companions and I parted; in a few minutes of warmth and formality, the night's climb was over. They, my teachers, were going down a trail, and I was going down the ashy side. I had no idea. Down the cone lay a field of ash. It was firm enough to walk on, soft enough to tumble on. There was not another human anywhere I looked. No vegetation. Thousands of feet of ash spread below.

Perhaps it was because walking down steep slopes is difficult, even laughable in ash. I stopped a few seconds, threw the staff down a bit, and dove after it. Only a short leap, and I landed in a fluffy bounce. Farther down, I picked up the staff. I threw it as far as I could, ran, and jumped after it, not caring how far. That one took a little wind out, and I sat a bit to get it back. Then all the way down the slope I threw and ran and leaped and rolled. I did not care about the slant of the earth, or about up or down or being black with ash from boots to hair.

Before Fujiyama, the climbing I had done was much a matter of balance, of feeling where my body was and where it would go. Here, coming up in the dark, leaning in and out, I felt the steepness more than the balance, in the feet and ears and staff. Then the earth turned to its light, and in that clear, sharp air of morning, my eyes joined the other senses. The space was found again. Now, flying and tumbling down the mountain, I learned a refinement of space by losing it altogether. In those hours down, leaping and rolling, I was only doing that, keeping up with the flying staff, flopping down the mountain.

The base came too soon. When I ran out of ash, I sat for an hour, bathed in flying. Looking up the slope, I saw very little. Everything was now above the clouds, over them, and then down, beneath them. I got up and walked for the train station.

I looked again back up. There was nothing up there but the mountain and the people, beyond where I could see, above the clouds. No other sight. And I was back down. No one can do that without the images staying forever.

That space of difference one learns in Japan, the creation of real illusions, must come in part from Fujiyama. The canvas of all space is up there. It can be filled, created, played with, moved around, set as the merest and grandest stage. The stacked bowls. The empty, linear room. The shoji screen. Apertures of any dream at all; lenses of the real. They are up there, and down again from Fujiyama.

6. Walking The Sea

There was nothing there, nothing I had come for, and I did not know what that was, either. If I had had some idea, some image, a notion, it had gone under ice or fled over tundra as endless as my senses were numb. There was just nothing there.

Despair has hope as an ancestor. The alien is searching, even without a destination. Neither hope nor search had brought me to Point Barrow. There had been a certain curiosity, but at that moment I could not recall it. Everything was cut off, missing.

The sea spurted laps at the shore of gravelly sand; but the Arctic Ocean was all ice, past the horizon. Turning around, facing land, everything was tundra, lifeless it seemed. A few quonset huts arched over the land with a kind of hostile stolidity. I had gone to the shore to get away from them, having checked in with the Navy people in charge after the plane landed. I had found my cubicle, dumped my gear, and headed, fleeing, toward the sea—where I found that Point Bleak would have been a better name, except that it would have been the name for every inch of that Arctic shore.

Sun-drenched lands catch the eye for their bounty, what is given. That wretched place was stunning for what was denied. It did not occur to me to wonder if the DC-3 which brought me had already taken off again for Fairbanks. It did not matter. Everything was missing. I walked along in the terrible crystals of sea and sand, fully aware that nothing would ever change again. I might have just walked the sea, where I would have seen nothing then but a sameness so stark it was bland, without drama, without relief.

All the signs of place together and all the inner states of a human animal can blend or fight, jarring place or person—or both—

into shattered rubble. Or making home. Those first hours at Point Barrow were far from home.

Part of that distance between home and rubble was covered by the trim baldness of a Navy operation, in the cold and bare of bureaucratic oil barons without titles, perhaps outcasts, in a cold and barren place. Even Fairbanks had not been that bad. There, in what appeared to be a village on the way to becoming a city, I worked on the same project, called PET Four, in the offices of a civilian company. The idea was to find oil at Point Barrow so Navy ships could refuel in the Arctic Ocean. It only seemed odd to me later, when I learned that ships could only come in two weeks out of the year. As the summer's exploration wound down, the company I worked for wanted to close a warehouse at Barrow; inventory, packing, shipping. I was happy to go exploring. Until I got there. Even at the end of the earth, anything run by the military has that same special quality. If life is not being taken, it is being denied. In the same way, what we think of as ambience, as environment, as place, is destroyed or denied. Getting the job done. At Barrow, one did not expect luxury, but the kind of eternal imagination which pops up like grass on a roadway was also not there. It seemed as if it never had been. Perhaps because we were a permanent gathering of transients, changing one by one, now and then. The military, faced with no evidence of a sense of place, usually tries to make up for it with bad food, bad beer, and bad movies. Just cart it out there, and they'll like anything they get. The great sadness is that it is most often true. Newcomers to such areas feel like those with eyes in the country of the blind.

Here, though, no one stayed short enough to grit the teeth and hang on, and no one stayed long enough to find the hidden life that waits everywhere for our attention. It was the interim stay, sighing into bland time to fill, no longer waiting or expecting, not discovering.

In the same way that I escaped to the shore after the first few hours, I escaped to the routine for the first few days. Up for breakfast; begin naming everything in the warehouse; lunch; naming; dinner. Everyone drank terrible coffee all day. Get the job done. There were five Eskimos working in the warehouse, with Clyde the head man and genius fixer of all things. It all began with Clyde.

During breaks and in off moments, he would teach me words in Eskimo, as I thought of it then. That led to other things about their life, their land, the ocean, how things had changed since the Navy came. He told stories about ptarmigans, polar bears, the berries of the tundra, the small life in the covering growth. Eventually, he asked me to his home among the tents of the village which was the summer site of the Eskimos. And I could not go. The village was off limits. Venereal disease, given the Eskimos by the whites, was so rampant—the story went—that we were not allowed. That was probably my first fury at the stupidity and inhumanity of governments. I directed that fury against the most immediate purveyors of idiocy (who of course went to the village at will), who responded with the usual simple-minded expedient. They tried to ship me back to Fairbanks. Oddly, the company I worked for insisted that I stay. From then on, I ignored the Navy as much as possible. Now, I would go the village. Then, I did not.

So I walked the sea instead. There was always daylight in summer, but it snowed on July 27th. Such things, sprung times for a new-come man, simply had to be attended in the way that one drifts unresisting into enormous revelations, as well as into disasters.

While I cannot remember ever hearing it as a child, I suppose we learn in many osmotic moments that the drama in our lives comes from hyperbole—the high mountain, the endless sea, a winning, grief and pain, a small sadness unforgotten. Those events "out there." Our very own way of looking makes them grand or mean, chosen or forced. One must learn the rare art of subtlety, for those arts depend on the casual, ignorant moment.

It was that suspension of hyperbole, a neutrality, that ignorant foraging with nothing there, that made Barrow an inexplicable and indelible place. Clyde and his friends and I would walk the shore, and they spoke of winter. But it was not as one might speak of fallow, resting land, a waiting. It was merely that a season, too long without an interim respite, would come upon them next, as always. We went out in their boats near the tenuous ice floes and we joked about how sorry they were they could not take me for a look at the North Pole just then. It was right over there, but one did not want to disturb polar bears. We had gatherings on the beach for all, and I was finally there, in Barrow.

Alone, I threaded the tundra, squishing and looking. The top layers melt in summer, making bogs of tiny plants and mounds of petite islands, and one should never even walk it, much less drive a Weasel over it. At the age of twenty, I should have known that, but in 1952, the Explorers Club mentality took small numbers of people everywhere and told the rest of us why we ought to go if we could. Today, the Sierra Club mentality takes hundreds on hikes and trips and tells them and us why no one should go there. But even as chewed up as the tundra is today, we give it much by staying away entirely, alone or in our ubiquitous groups, which we shall not do. Man is an animal that travels.

Nothing like that even occurred to me as I clambered the tundra with unbelieving eyes. I began to see that it was an absolutely exotic place, which only means that it was a rare and exciting place to an alien. Even Clyde did not know some of the things I described. They were of no use to anyone.

On the shore, I looked down at the stones and wondered aloud and smoothed them off and peered and dropped them back. And I looked out at the ice pack, keening for the Pole with just my ears. The waves were very slack. I looked out at the sea; to my left and right; turned around. I fixed the land, flat as seas. Even mountain climbers meet their match.

My inner ear heard a chortle as I walked back to the quonset with the cubicle with the cot. Because I thought I might take another look at Kansas. Alaska was almost over.

The Navy ships came in August, bringing supplies for those who would stay the winter, but taking on nothing in the category of fuel. It was a big day for everyone. We were all rescued without being saved. The ships left. I got ready to leave. I had my certificate for The Top of The World Club, having crossed the Arctic Circle, and I had the famous PET Four parka. Hundreds of crates from the warehouse would follow me. There was no one there to say goodbye to, or to thank, but Clyde and I had muttered long visions we understood. The plane took off, and I looked for the Pole.

Back in Fairbanks, I reported in, back among my colleagues. They were curious about how it was up there—the food, the boredom, the Eskimo women. They noted my silence and credited it to shyness and the wisdom that youth notices nothing, and we were all working away again.

My immediate boss was Saad Assad, who claimed he was the world's farthest north Arab. As the crates from the warehouse came in, he asked me where the inventory list was. In seconds, I was numb and speechless. I had to tell him that both copies were packed in one of the crates, but I did not know which one. I thought he might tell me the truth, that the tundra and the Arctic Ocean had addled my brain—in ways he could never know. He only looked a little stunned, then bent his head, held on to a drafting table—and laughed wildly. He knew. After the laughter, he invited me out to lunch and asked if I would stay instead of going back to school.

But I had already known I was ignorant, and that summer filled me with the taste which comes from knowing that. Up there on the Arctic, I had learned from the people, from the work. There was a kind of continuity about that. Just knowing more things carries over, informs the next day, whittles on one's dumbness. Clyde. Saad Assad. Point Barrow. All the PET Four nonsense.

But one cannot have strong images of nothing, which that land seemed at first. We attend forgettable things only if they are tedious. What stayed with me was odd, it seems now. The image of emptiness filled me. It turned me from the Arctic to the tundra. The empty bleakness led nowhere, everywhere, was itself a place I was. On land, I had nothing but signless horizons.

I remember that, see it. Yet there is an inescapable image to follow, another continuity which always comes with looking out at sea and tundra. I look down. Shore pebbles. Inland, mire. I move, and place myself.

Those bare few moments of attending the body in space depend on quick perceptions of place and of movement. It is in the small, directly sensed, precise moment. The animal instant. Riding the earth, attending earth-set, walking the cone of Fuji-san, knowing the richly barren land of the Arctic all came back to me, left without invitation by the gifting birds. Those morsels went together on the same shelf, but with only a vague reason, until later.

That sense of our bodies, our selves, moving in space, placed, is not just "out there," where we refer, where we attend. It is also "in here," where we are sometimes unable or unwilling to refer. And that sense of our bodies in space is nourished by what might be called betweenness. There is space or some object between me and every occasion of all my senses. It is betweenness by which I can parse every singularity. And it is movement in space by which I know the actual experience we call time, and see that a moment and an eternity are the same. They are ideas about the measurement of time, and we never know them in our senses. Our only experience of time is that of motion. We are literally moving among movements, changing among changes.

One of the most intense experiences of Trigant Burrow's notion of cotention is that animal instant in which we suddenly recognize the real, literal motions of our world. A certain movement and a certain stillness make us attend our perceptions more than the idea of time does. We first see the fullness of our place, the richness of our animality, in those quick attentions to the moving and the still. Yet stillness is a story, a myth made of movement. What seems still makes us attend movement. And therefore space.

The room around us overflows with more than we can translate through our senses. It is our place, so full we cannot notice it all. In the animal instant, we recognize our selves, here, in this place, or, as Heidegger said it, "here rather than there, now, rather than then." In the sudden flicker of knowing both Fujiyama and bonsai, awash in the sense of ceaseless motion of my own and of my place the earth, I reap seconds of myself in a world with no thought between us. Not oneness. Not separateness. An untranslated sense of where I move, it is unconnected, unspecified, but more definite than stone. That quick perception of moving and residing has the intensity and vividness of an esthetic perception, as if standing with a dumb stare in the presence of cobalt blue, or with an inarticulate ear washed by Beethoven. Later we can speak of spectrums and semihemidemiquavers. Later we can speak of the nature of time and space. We can connect our sudden, isolated sense to a whole life. Or not.

When I begin to attend those instances of space and moment, as a human animal, I come upon others. I glimpse other lives, gather quick or long days attending those creatures of whom all, some, and none are like I am. Yet in the human niche, those sensory moments and then reflected knowledge are both simpler and more confused than with a sense of time or space. Knowing a little of ourselves, we know a little of that other one, who still remains entirely other.

IV

IN THE HUMAN NICHE

Perceptions as intense and personal as a moment in time and a feel, an orientation, for where you are in an immediate space are not to be shared in any precise way. There are recognitions and attentions. To see what one has seen, to attend to what one has neglected, can be surprises picked up here and there, such as those left to me by the gifting birds.

I thought for a long time that what a sense of place must be had to do with just the lay of the land and the look of it. Mountains or prairies, a certain hollow or a rocky coast. But I had from the gifting birds a shelf full of people and manmade things like stores and cities and airplanes. It is quite clear that a place to be yours has its people, has the imprint of man. Place is not made just by time or space or esthetics or the earth or lives shared with other people, other animals. We come to each other one at a time. Man shares that with all other animals. We must be among each other, and even the hermit and the recluse have known days among the others.

Surely there must be people made by places and places made by people. Except that is not often so. There is more balance than that, without parts to isolate, without pieces to pick apart, without man apart. Architecturally, a niche has its wall and its building and its exterior. Ecologically speaking, a niche defines an organism by the place it has in terms of its own characteristics and its relationship to other organisms in other niches.

A place is not just its physical offerings, not just the hills of San Francisco, the barns of a farm, the snow around a tarn in spring. That is no news. Yet there is a freshness to see people bound to a place, place to people, and watch them support each other, do fine or terrible things to each other, and see the indelible trace running through people to place, and back. At times, for me, the connection of person and place has been so intense that the place should not be seen again, like a face left unrecalled.

And there are other times when you know what a place is not. I found that in the vapid ambience, the severe humanness, of a jetliner. There, the people seem no longer to come one by one, but in a styrofoam package, where a person is disconnected.

Perhaps there is no bond to a place through someone else as strong as there is in loving, or even in knowing a touch is a passing closeness. I am such a cynic that I do not even believe my own cynicism. Yet I learned in Mexico at an early age the bond between quiet, soft desires and the edges on which they are cut, just short of vanishing, just this side of never returning. That balance can be learned anywhere, but for me it was in a dance hall in Mexico, which I was to remember by sight and sound and smell as a lesson in the possibilities of the quick heart.

7. The San Gregorio Store

Elderly she rises there above the flood plain of San Gregorio Creek. Since she has been there, her skirts have been dry. At one time, closer to the creek, she was up to her knees in water every winter. Her appearance is deceptive, though, for she is a building only some fifty years old, anchored like a ship eternal, going back fewer years than her look, and forward not at all. For which we are thankful.

At the front of the building there are three lanterns in arched niches at the top. Below, the windows are arched stucco with colored tiles, and the false-front roof is aged orange tile. Certain architects might guess late twenties or early thirties, and be right if they chose 1930. That was the year Peterson and Alsford, General Merchandise, was rebuilt. It had been there for fifty years before fire, that great shaper of small histories, had burned it away. It was not the first removal, for the place was a successor to Levy's and The People's Store down on the banks of the creek. The fire was, though, the last removal of the old records of the Wells Fargo stage which stopped on a daily run from San Francisco on its way south to Pescadero for the night, and the return north the next day. During the fire of 1930, enough neighbors turned out to save the owners' houses and furniture, including pianos moved away from fire's harm.

Today, though, it will surprise you, this store. Even if you were good all year long, Christmas would not treat you so well. On your left as you enter, the stove, the tables and chairs, the piano, the rows of books all request that you sit and hunker for the winter, with just the right drink in hand. So. To your right at the entrance is, could it be, a bar? Get anything you want, as long as it is a drink, not a cocktail.

You may survey the stoves on the floor. They are the cast-iron eternals, which fade every year in favor of burn-through-in-your-lifetime-but-hot models. The supply is in part a matter of what is to be got from oldtime hardware wholesalers slowly going out of business. There are just not enough lovers of cast iron around anymore.

Here at the bar, you may find, at nine or ten in the morning of a weekday, a few San Gregorians picking up the newspaper and having an early beer or brandy. If enough are there, a quorum is declared and business begins. It concerns the day's news, the latest horse show, the state of socialism in the fifties, the principles of weather—especially of fog—and what is growing well in the categories of chokes, sprouts, beans, and pumpkins. At times, if its members are in attendance, there will be an impromptu meeting of the San Gregorio Philosophical Society, which discusses the infinite verities of the universe, even if there are none. Mike, once in charge of a cooperative worm farm, believes that everything happens according to a celestial plan. Charlie, a man given to horses, thinks some things happen that way, while I see only chance everywhere. We debate this question as if we were School-men of long ago speaking seriously of angels dancing on the point of a pin. It only shows that almost anything is tolerated at the store.

Beyond the bar, the food begins, right and left. My taste buds lead me always to the big glass cheese box filled with jack, teleme, cheddar, swiss, bleu, all blocks and bricks and rounds. Once the box opens, there is no retreat. The salivaries begin. Around the cheeses, fruit and links of salami, sourdough bread, jars and jars of pickles and peppers, pig's feet and relish, lupini beans and menudo, all keyed to resident and passing gourmets.

Just a few years ago, Eric was behind the counter, usually tending bar and telling one tale or another about the place, the people, his days in France and England. You would have guessed he was surely young to have been in World War I, since he looked about 65 or 70. But no, he told the truth then, at 84, working on his feet every day. His great joy in life was simply people. He would ask where you were from, what you did, introduce you to the "kids" next to you, ask just exactly how you wanted your drink.

If you look out from behind the bar beyond the stoves and sweets and food, above the oiled wood floors, the ceiling towers over you, covered, as the walls are, with the patina of old ivory. Ancient ads are tacked over the paint. They speak for Lee and Levi, for Fuller Paints. In the paint ad, a boy sits next to his dog. For some reason, the boy is speaking into a can-and-string phone and he is saying, "Yes, ma-am, they last." Under these signs, the clothes and boots and utensils await the patient and gleeful gaze of everybody with a feel for oldtime reality. Exquisite kerosene lanterns sit next to a butter churn on a shelf above sturdy stoneware, 20-cup campfire coffee pots, water ladles, flyswatters made of metal screen, cast-iron cookware, and suspenders.

Buying or not, this is a wanderer's store. As in hunting mush-rooms, you must get an eye for a certain kind of seeing, in this case for corners and odd crevices. Such an eye just may turn up, say, a fedora from the forties, still in its box; a special kind of thread not made anymore; or a shirt ten years old, still in its pins. Barefoot young and corseted old folk stalk the aisles, where they may read, "Please, no food or drinks in this area," "Please, do not unpin shirts," and "Bare feet at your own risk." This, the clothes department, is carefully watched for flatlanders who would shoplift, like the man who left his boots in a new-boot box and wore the new ones outside—but not far. He had not counted on the eyes of Beth, Eric's wife; Nancy, their daughter; and Bob and Hazel, who are the Petersons. Of course the store is known simply as the San Gregorio Store, for there is now no other one there. Across the street there is another gas station and saloon, as well as an antique hotel, no longer receiving guests but once filled in summers. But back to the man in the new boots. When approached on the way to his car, he said he was only trying to see what color the boots were in the sun. He was escorted back inside where he tried on his old boots and was asked to find out how they looked in the sun.

Weekends at the store are madness, and some locals will not go, not after noon. The main aisle can look like Wall Street at lunch hour, a mass of moving people flowing in and out. Petersons and Alsfords were everywhere, waiting on, watching, and smiling at the crowd, which never seems to be noisy but always in a hurry. A solid peace is kept, always.

Today, the peace, the wares and fares are all much the same, still here. But Eric is not, or Beth, or Bob, gone from the earth. The others live nearby, having sold the store. Everyone wondered about the new owners, the new changes. A long time ago there were dances, the bar was open at night, there was community. Would the new people have that again? Would it all go modern? Synthetic cheese, car parts, acrylic undies? No, but there were changes upon changes. George and Joey, Clay and Ellen, and James bought the place, and what made them like it was the way it was. So they made it more that way with good, oldstyle things, added the table and chairs, the piano, the great iron stove. The folk sighed. The world had not gone mad. Here.

There was more to come. The partnership did not work, and it somehow came out that George and Joey stayed. Their eye is right; they fit and feel it. And they have their hand in more than being merchants. Joey is an attorney, though she does not practice law now. George still teaches a little philosophy at Stanford. Got a few head on the pasture by the store. It is good security for us, the habitués, that they are not pure-bred merchants. They sense the history of the place, hear the walls, imagine the character, shape it gently. It is now their niche. It is thought that we have all imprinted each other with the solace of shared lives. It is more than good fortune that lured George and Joey. The place, the people here, were hoping they would show up.

For those who stop at the store and can see, the obvious bounty of the place is its kind of peace and its ambience of age. For those of us who live near it, there is now even more to celebrate in its present and future. After all, here is a store where people are living a way of life, not just working at jobs. Here is an actual place where you can also buy clothes that last, basic food, hardware that never wears out, equipment that works, a good and honest drink of whiskey, all in the same place.

8. Up In The Styrofoam

Every place on earth is a moving one, from the flicks of the smallest microscopic world to the inching of massive mountains in the great ranges. Stationary things are still moving with the earth. It is difficult to sense that, but possible, especially on ocean or desert. Necessarily, your senses must understand that at sunset or sunrise they are grasping an important reference point. It is easier to see, in those moments before dark and light, this essential act: The earth is turning away from or toward the sun, and the daily fire of our lives is not setting or rising at all. I find that lying down is the best way to see it, to sense at sunset that I am moving away from the sun. Except for night-blooming cereus, it is even more difficult to sense the slow movement of plants.

The faster movements of our places on earth are truisms of each day. The car, plane, train, the lesser known spacecraft's quickness. We may spend hours there or weeks, but they are places in our lives as surely as forests and cities are. The train is my favorite. It has marvelous amenities like food and drink, a speed and timing suited to seeing the earth at a close range, and a good survival rate in case of accident. Planes are more curious.

My first plane ride was at the age of nine in the summer of 1941 on a DC-3 from Dallas to Mexico City. Though I was very excited, I became convinced before we got there that one could die of airsickness. But it did not prejudice me against planes, and it never happened again. In fact, since then I have ridden many miles in a DC-3 and have fond memories of it the way my father did for the Model T. First, you could tell the plane was being *flown* and not just watched over by people in a cockpit or flight deck. It was responsive, too, and needed

very little room to move around in, so that made you feel even more
that the hand of man was moving it. Your ears and stomach knew, too,
that there were engines out there, and their affairs were not subtle. On
the commercial model, the upholstery was cloth, a kind of wool
corduroy. I never sat in one of those seats nude, but I can imagine it
would have been like wearing a hairshirt all down your backside. The
cloth was rugged, though, and the feel of it at least genuine, along with
the smell. The cabin smells were as real: tiny air jets exuded metallic
air; the residues of cleanser, tobacco, food; the mingling of perfumes.

Looking back, it seems that even the passengers were different,
and the stewardesses were an earlier and separate subspecies from
flight attendants. Then, they were all what one might have called
lovely and gracious and warmly personal, and that was good if you like
that sort of thing, along with more drama. In 1952, I rode a DC-3 from
Seattle to Fairbanks. There were not many people aboard, and the
stewardess and I started talking. She was working the Alaska run, it
turned out, to escape. Her maniacal boyfriend had followed her from
one city and airline to the next, always threatening to kill her if she did
not say yes to his proposal. She thought perhaps this time he would
lose track of her in the frozen north. I guess he did; by the time I
returned four months later, she was unfound but not lost. On another
flight, from California to Colorado, I had just had a spine operation.
The DC-3 was not always the best at maintaining air pressure, and it
began to hurt my neck. There was nothing for it—no booze, no pills.
The stewardess came up to my seat and said softly that if I would not be
offended, she could give me a couple of pills she used for pain. I
accepted, and they helped. Midol has more than one use.

Long flights in a DC-3 or a DC-6 could seem twice their duration
in your head. Endless hours. Once, sitting next to a geologist, I got a
perfect lecture from Denver to San Francisco on how mountains were
formed in the West, by a process more horst than graben, with
continuous samples flowing by below. As a captive and fascinated
audience, I remember more of it than any college lecture. About
another flight I remember less. Coming from Honolulu to San
Francisco—a nine-hour trip—the crew on the non-scheduled airline
discovered that somehow no food had been put in the galley. Nine
hours. No food. However, there happened to be an inordinate
number of cases of champagne aboard, and plenty of paper cups. A

bewildered ground crew in San Francisco poured a load of passengers *off* the plane. The prop set was not so bad, either.

When I think of jets as a place where a lot of people spend a lot of hours, I wonder about our changing sense of place. The jet violates my sense of place. Take the smells. The air, the seats, the cabin aroma, everything smells of plastic, has the feel of styrofoam, vinyl, formica. I ought to be glad, the way I ought to be glad if houses could be made—and they could be—of plastic. But I am unjoyful and unrepentant, for one reason. Plastic is catching. Since there are not enough lovely and gracious stewardesses to go around anymore, and no one would notice anyway, they have a facile charm from the same mold. Notice the passengers. They are suspended: in mid-mission as businesspersons, on vacation and holiday errands, sitting with apparent pleasure in an ambience just like home, office, and cocktail lounge. They are not flying, they are moving. They are not in motion, they are changing place, in a place like all the others. On edge though. Some show it, some do not. But plastic is catching. Families bicker, the body shop is still open. What everyone wants. In a way, moving through the air in a jet is like hitchhiking. Unknowns meet for a brief time when anything can be said, pretended; any story told; any lie harmless. The place has been built for it.

A lot of people moan about plastic, and it is at least trite, no matter how true, to speak of the plastic smile, plastic food. Here there is something added. Up in the styrofoam, the place is like many others, but is different by danger. No matter how much anyone "loves to fly," only an idiot would ignore that. The fear is real. Like everyone else, I have heard the litany: Flying is safer than driving a car; stretched over millions of passenger miles, the accident rate is low; it is only a chance in a million that my flight will have an accident. Folly.

The mortal reality is that a modern plane is plastic to the core, coated with acrylic. No one notices. All is accepted, marvelous, abundant with laughter. Contrariness like mine gets a patronizing grin or comments about new times and a modern age.

Such reactions are irrelevant. Though I get plane-sick now rather than airsick, I have nothing against old, new, or future times. What I see in the jet is a sense of place gone sour, lost. Both the builders and the buyers—as in rows of identical tract houses—have conspired to deprive themselves, witlessly and literally, of their senses—the sight, sound, smell, taste, and touch of small treasures vaguely remembered. And gone forever. I am realist enough not to mourn anything gone forever. It is only that I keep thinking my senses are real, and I miss one of the places where they used to browse.

9. City Of The Senses

If there are any confessions I should make about cities I know well, there would be two at least. Dallas is unsufferable, and I cannot take being very far away from San Francisco. The latter prejudice, shared by many, is simply a matter of love, unqualified and unrepentant, among all the realities there, which are no better or worse than in other cities. Though I could not live there or in any city, I live in the country near San Francisco on purpose. To get a balanced environmental diet, I go there about once a week from my home in Pescadero.

One hears many phrases about San Francisco—Baghdad-by-the-Bay, "pearl gray city of love," gourmet's Mecca, and many about what appears to some to be a whimsical population. Someone once said the perfect symbol for the city would be a miner Forty-niner in drag. And if New York is the Big Apple and Los Angeles the Big Orange, surely San Francisco is the Big Lotus, city of dream eaters. Which is how it all got started. One hears constantly from those who will probably never know any better that the city is the West Coast center of lunacy and perversion and evil, an attitude shared and spoken by many an upright, loving Christian. But San Francisco's beginning must be remembered if the city is to be understood.

The place as city was born with practically no gestation period just a little over 125 years ago. It sprang live from the head of golden dreams. People from literally all over the world arrived by the thousands. Whole shipsfull—passengers, crew, captain—merely stopped in the bay and fled to the gold fields, leaving hundreds of derelict ships behind. In time, these flood tides had their ebbs, and the seekers came and went, always passing through San Francisco, dreams alive or in tatters. Those who supplied goods and services to the fools

65

and wizards stayed and made the durable money, became as dull as the socially elite everywhere. But the wanderer, those on the roller coaster of rags and riches, the odd ball, was tolerated, welcomed (he was buying), and then became the object of a kind of embarrassed civic pride, an amusing attraction. And we are now only in the fourth generation away from those Gold Rush times. No city can be expected to take the step backward into the Puritan ethic overnight.

To me, though, this city is one dedicated to a manic choice among the senses, the five or six or more of them both sensuous and sensual. If it is a city person's city, it also belongs to country people. At least I see it that way, sensing it with an attitude similar to or perhaps the same as the one with which I sense the redwoods, the mountains, the ocean of my home. Each part means something to the other. And after all, the city is an environment natural to man, no matter what harm may come to him by creating it. Its life is still entire, though no other city but this one has that effect on me.

And yet every time I go there, my senses tend to fall apart. Going in from the south on the coast, you are first confronted with Daly City, that town which of all others best fulfills the American Dream. All the houses are the same, row upon row. In each picture window, a lamp.

Depending on the route, one may then spy Mount Davidson, highest point in San Francisco, the city some say is the most sophisticated in the country. Certainly the art galleries do not speak to that, and this hill does not. On top of Mount Davidson is a grotesque concrete cross over a hundred feet tall. Here, for those who can stand the sight of that cross and a sunrise together, Easter sunrise services are held.

And then, in the city itself, are the lovely, lissome Victorian houses painted with colors only a city with imagination could brush on for the eye; hotels like pleasure domes; the great cookery and hardware shops of Chinatown and Japantown; North Beach, the Italian-bohemian-beat-hip and sturdy family neighborhood; the separate characters of each district.

To keep all this together in one city of so small a size takes a nimble eye, perhaps a city eye which I do not have anymore. Somehow, when the fog fumbles its way over the city, covers it all close down, these different visions all make perfect sense to my eyes. Each sight keeping to itself, they still fit together, bound by fog into visions which can no longer be seen apart.

Wandering around town on errands of work or entertainment, one hears that the sounds are special to the city. New York City seems to have no sounds, only noise; but a catalog could be done of San Francisco's sounds. Cable car bells thud, tink, or ring, but always with a melody by the gripman, a musician, at the controls. The throaty basso of ships in the bay seems to be in the air any time you want to pause an ear to keen for it. If it is close enough, say at dockside, that bellow can go right to the stomach. These voices and others, like the cadences of many languages, cling to the sharp air of a sunny day as if they were seasonal winds expected by everyone. Perhaps if these sounds were not in the air residents would feel the same nervousness the people of Hawaii feel when the trade winds cease and the air itself loses its rhythm. A foggy day is another thing. Sounds seem wrapped in gray, anechoic winding sheets, dead, going no farther than their sources.

One may actually choose a fragrance in San Francisco, from coffee roasting, through fish, the sea, bus fumes (some feel homesick without them), Chinese food, mock orange, the aroma of new buildings (which come in endless waves, like the fog) and old. Many cities have a smell, one smell, a blend of the best and worst. High air with mining smells and diners in Butte, Montana; the ancient odor of iodine from the sea with cotton candy from the the boardwalk in Santa Cruz, California; crystalline snow and rampant smog in Boulder, Colorado, all swirled together. Not so here. Perhaps the winds will not allow these eddies of the nose to blend, or maybe the hills hold them home in their native valleys. In San Francisco one may find a way to food coveted by the tongue by simply letting the nose test the ambience of a place. To do that, one must not merely smell, but relate the aromas, catching quality of food, place, the lambent signs of a personal empiricism. Of course, like every other city, San Francisco can accommodate anyone whose urge is strong for good, solid junk food.

The separateness, the well-defined character of every sensation keeps to itself, puts my senses back together. Something else puts them together, and that is choice, the liberator, the teacher to reason or whim—or sight or smell. It is a delightful feeling to have many possibilities informing an active sense of choice. Thus many people, even some of those who live there, love San Francisco as a place, not as a mythical kingdom.

No matter how much I can come to know my feelings about this city with its strong sense of community, there is always one of its promises left over in a place alone. For everyone with any city there are a few or even just one of the faces on the frieze which mean more than others. For me, here, it is the hills. I do not know or care if it is simple or simple-minded, but those hills are the face of the city to me. They give me renewable choices for the esthetics of the eye and the inner ear.

In any countryside where I lived, I would always want both oceans and mountains near a fine city. The geography and topography of Kansas are my idea of madness, not to be indulged in very often. Maine bathes in mountains and oceans as much as dreams could say but has no city. New Orleans is in some ways a fine city and has ocean nearby but is as flat as a swamp.

San Francisco, with bay aplenty and a little ocean, is like other cities by the sea except that oddly enough shipping is not its passion, giving it an inland flavor it earned in Gold Rush days when those people came from all over the world just to flee inland, leaving a graveyard of ships. Leaving other minds to cherish its hills, steep and full of more than views beyond the mere lights of a city. Leaving a city with an entire table of contents for the senses. Leaving a city with a sense of place.

10. El Salon Mexico

Around me I saw nothing but flashes of lights and darks, the men in white shirts, the women in dark shawls, all at tables or dancing, moving somewhere, somehow. The ceiling was not just high but enormous. Music. Everywhere. All over me. It was like this as I look back from now. Today, loud driving music gets to my stomach. The music of El Salon Mexico got into my feet, in the calves of my legs, up in the pelvis. I stood there at first like a stone, trying to awaken under the riving of roots.

That moment, that night, in Mexico City stays inside like bone. I was alone there, 16, and those first seconds gave me a feeling I had never known. Earlier, there had been other and quieter times when a place had bound me to it, but over a longer period. In a cluttered Texas garage back of our house I had gone among relics like any fevered young seeker would, and I discovered music. An old wind-up record player with one record. I played it so many times I hope never to hear *Humoresque* again. In another house, there were woods behind us, and I became forever addicted to rabbits, scorpions, and tadpoles— and I learned that inch-long frogs, like most other animals, die if they become just pets. The Texas places were ones where I grew up in and around family houses. Though we have no choice in where we grow up, we learn it in time.

Mexico was the first of many real and chosen dreams which have passed instantly through my senses and never left. I cannot remember or trace the very first days there, for I was nine. Mexico City comes back from that time like a candle, with flashes of clear light such as the rush of buses and cars, and with hinted shadows like the tastes of street food and the smell of burning wood. There were several summers between that first one and the one of my sixteenth year, spent there alone. Those summers had sudden moments which were the result of a long kinship. While riding on a bus one day, I read a billboard in Spanish and realized I was not translating it, that I knew the language. After perhaps three summers, one Sunday I was carried away by the bullfight, cheering for a good pass, tingling in the spine, which I was not to understand for years—if ever. On the third time riding a horse to see the volcano Paricutin on an ashen plain where the lava flow had left only the church tower of a town, I first felt an affection for rocks and other mysteries.

Short perceptions, bare seconds, stay, too: the taste of a mango; the first look at the calendar stone and a pyramid, both more ancient than I could understand by their ages; the human ruins on city streets, like nightmares, barely seen and with no understanding, yet still there now, recorded.

All of those things behind in my mind, I stood at the portal of El Salon Mexico looking like what I was, a tall, skinny kid of a gringo with a face that could have been gazing on Taj Mahal instead of into a dance hall. The talc-white shirts belonged to men who did what they could—sweep up, forage, steal—for the life of a city and for their own. The shyly wrapped shawls modestly covered mostly *criadas*, the city's maids; many of them were very young. Everyone danced as if they were wooden or on their way to hell and didn't care. Stepping in, neither did I. That enormous ceiling covered a U-shaped room with bands going at each end and in the middle, all playing different tunes—Augustin Lara, mariachis, rancheros, La Bamba. A small polyp of people around each band danced to the music played; the rest moved as they felt.

First, I wandered through it all, with polite footwork. There were tables along the walls, and men and women left and came back like workers to and fro at a hive, though this was a worker's holiday. Then, I spoke Spanish as well as English, but I was shy and aware that I was not part of the holiday, and could not be. I was partly wrong.

At a table near the door, I sat down. No one noticed or said anything, and that was a relief. Earlier that summer I had ridden the hills around Oaxaca for a month on a horse. At first I seemed like me to me, but it did not take long to understand that a kid-gringo in those villages was a curiosity. The interest took the form of invitations to eat and sleep in homes along the way. We shared stories and tobacco, talked in the night, and my hosts joked slantingly about where and with whom I would sleep. Such attention was new and I had to bury my shyness in order not to offend anyone. But at El Salon Mexico, where gringos often came the way they went to Harlem in New York for the sights, no one seemed to notice. I watched and felt the rumble of the dance, watched the tables fill and empty, watched the bottles of tequila and beer tip up, watched the sound going into people's legs.

The air smelled of Delicados, those wonderful Mexican smokes with the force and acridity of Gaulois, a French cousin. I stopped watching and looked over into the semi-darkness of the wall behind the table. There must have been someone looking back at me, or was there? I saw her there, anyway. I leaned over the table a little, and yes there was. She was no older than I, and we stared at each other like ancients on some known ground. After looking away and back a few times, I stood and put out my hand; she rose and took it. At first she looked down; I spoke to the top of her head; she looked up, answered, and did not look down again. We talked, as learners do. Her name was Amelia, and she lived and worked as a maid in the area of the city where the streets are named after philosophers. We danced.

Going back to where she sat, there were more people with us, her friends. Those who did not giggle, grinned, and suddenly we were lovers. We giggled and grinned back. They joshed us on, but we did not need it, looking at each other, saying little, and noticing her friends now and then, as asides. All of us loved it, and I felt less and less like shuffling my feet or biting the end of my collar. We danced, stood, sat, looked at each other, said nothing more about anything outside of El Salon Mexico.

It was early morning when her friends got up to go. Amelia rose with them, the music still pumping. Before I could speak, they all said goodbye. Amelia and I held hands, she looked at me like a child in trouble, and they were gone. For a few moments I did not know what to believe—that she had been there or that she had gone. Sitting and watching again, everything looked as it had before I had seen the dim form in the shawl, except it seemed the music and the dancers had slowed down, the lights brightened.

Twice I went back looking for Amelia, but never saw her again. The second time, walking home in the rain, I realized what Amelia had probably sensed in extremes of grace and sorrow all her life. The kind of intensity we had imagined or known is only of a time and place, ours that one night in El Salon Mexico. As I walked, I felt it in the excellent rain.

11. Colonel Wesson's Handgun Lesson

Almost daily there are signs of our society's obsession with the use of the handgun—presidents, popes, dealers, children, aunts, uncles dead by some quirk or plan. There is one handgun which seems to be at the pinnacle—though perhaps not the most common—of the romance of shooting people. It is the .357 Magnum.

Clint Eastwood made this cannon immortal in a series of bloody, intensely forgettable movies about the absolute force of man, Magnum, and machismo, with a character on his way to a dubious but definite personhood—as a killer. Which is a common, simple-minded way of exposing oneself as a man.

As these revulsions unfold, I keep remembering the man who invented the .357 Magnum, Colonel D.B. Wesson of Smith and Wesson. I knew only a brief period of his life, when I shared it for a few hours a week. I don't know, but this may be possible: perhaps he even hated guns. But he taught me how to shoot one, it could be said, in his rather odd way. He and his method were quite remarkable—and maybe hopeless.

I have a cousin who was, and may still be, just crazy about shooting. He was in the Air Force, stationed on one of those bombers which went, in 1951, up over Alaska or somewhere for lunch, and then came back to Tucson. When he wasn't eating lunch over Alaska, Sam was out on the desert. Shooting. One day he drove past a small sign which read, "Colonel D. B. Wesson." Sam wondered. He drove in, walked to the door, knocked. The door was opened by a slight, wiry man with slicked-back hair and little glasses, cigarette hurling smoke toward his eyes. My cousin said this: "Are you Colonel Wesson of Smith and Wesson?" And the man said this: "Come in, boy, come in!"

After three hours there, my cousin came just about leaping into my room to tell me that he was going to learn the Wesson Method from Wesson himself. He had included me in on the lessons. Now, I did not know much about handguns. "Indifferent" may be the best word. And I was a student at the University of Arizona, so there was the time problem. But Wesson wanted more than one student, perhaps because he had experienced some difficulty in getting his method accepted. In fact, his method had been called "insane."

At the time, the accepted method was to stand rigidly erect; raise the weapon up, pointed to the sky; lower it; aim and squeeze the trigger slowly with the hook of the index finger. Not Wesson.

Arriving for the first lesson, we found Wesson in his living room, a large pistol on the table in front of him. The smoke curled up from his usual cigarette. No matter how long they were, he always smoked as if the cigarette were a half inch from his finger or mouth, burning and filling his eyes and nose, giving him a chance to tilt his head and peer at you with a quizzical, unspecified command in his look.

He had a softly deep voice. Elbows on thighs, he spoke with a tone which knew much and gently insisted more. The first thing was: "Stand up now." We had to stand in a comfortable position, then close our eyes. Wesson had Sam open his and look at me. Then I looked at Sam with his eyes closed.

"You saw it?"

We both said we did.

Even if you stand in a comfortable position, when you close your eyes you begin to sway.

"Tonight, you will stay here until you find and know by feel, by instinct, how to stand with your eyes closed and not sway, ever." We left at midnight, with instructions to practice. And of course we did.

In the next session, we had to show we could do the stable stance at will. When he was satisfied, Wesson spoke. "Slump," he said. We looked at him dumb as guppies.

"Don't you understand what 'Slump' means? It means curl your spine down and stand there like everything your mother hated." Suddenly: "Slump, damn it, slump!"

Now he bellowed like a drill sergeant, and he did indeed have command. "We slumped, we hunkered down, and he kept at us. We had to unlearn all those commands to stand up straight. He got us to slump as if he had pushed on our heads, and he was about six inches shorter than either of us. We spent that evening learning to slump the big, the giant, the grand slump.

I was now curious enough to keep going, and we got along fine with Doug Wesson, but we never called him that. And we did not ask when we might be picking up the big pistol, still there on the table.

The next of the twice-a-week sessions we held the pistol. Just that. First I let it dangle by my side, then Sam. But we could only hold the grip between thumb and forefinger.

"Got that," he said, "got that pinch, like scissors? Don't ever assault a handgun with your grip, don't ever fist it. The bullet is the weapon. Now the scissors."

In the Wesson Method, you never hold the grip and crook your finger around the trigger. You cradle the piece in your palm so your thumb and forefinger push toward each other like scissors, with your finger almost straight in front of the trigger. There is no possible way you can tell when the thing is going to go off. And it never did, since it was not loaded, and we were always in Wesson's living room. The gun we used was, he told us, the prototype he had made for the .357 Magnum. It had an 8¾-inch barrel, and the front sight was filed flat on top. We would find out why later.

The next meeting was all about good sense.

"You have no idea how many fossils there are around," the Colonel said. "And they all teach the art of handgun shooting. Look here. The reason you held the piece up was to shake powder into the flashpan. Not needed now, but the fools still do it. You can't hit a thing between the sky and your target. Wasted."

So he taught us to let the pistol hang at the end of the hand, along the leg, and bring it up from out of your slump. "If you get nervous, you might still get a ricochet hit." That night, we stood, slumped, brought the piece up and scissored it off, getting the rhythm of everything together.

Wesson began to speak of graduation, and we had not fired even once. Graduation would be the time to fire, we knew.

One night when we were standing and slumping and scissoring,

Wesson was sitting and grinning and smoking. He said, "Time for graduation, gentlemen."

We looked with the same dumb look of the first slump session. "You first, Charles. Raise the piece, ready to scissor. Now...." He took a dime from his pocket and balanced it flat on the front sight.

"All right, now. It must stay." I scissored. The dime was still there after the hammer fell, but I was not. I was away on a moment's journey of disbelief. Then it came to me that I had done it: the hand, the hold, the scissor. But then it came quickly too that it was not just my graduation. The Wesson Method had passed once again.

On Sam's turn, he did it the first time, too, and then kept doing it. Every time he did it, it worked. I have never done it since.

Not long after that, in what seemed to be another life, I was walking guard duty along the fence of a supply dump in Korea. The fence was spotlighted and I had one clip of nine rounds for an M-1. We had been warned about people breaking in, stealing supplies. There I was in the light, with nine shots. I wondered what I would do. Finally I decided I would not try to shoot anyone. Not anyone, ever, for any reason. I do not know exactly why I made that choice at that moment.

It was a large moment. With just a mild background in hunting, and Colonel Wesson's handgun lessons, I guess I had thought about guns in the way a lot of people did. Learn them. Use them. Accept them. I think behind that, in some nether voice, we know what is unsaid: Kill. For that is simply what it is.

We accept that voice, spoken or not, in our society, and of course gun control is just the surface of the problem. The real trouble is that we supply—sell and make enormous money on—more small, large, and catastrophic arms than any other people in the world. If we had any desire to stop the killing, we would stop producing the means. The mere thought of that in the face of profits, our main goal in life, is a laughable silliness.

Which makes me think of two things about Colonel Wesson. One is a story I did not understand at first. Some friends and I were in the woods nearby a few years ago, and they were plinking with handguns. All they knew was that I would have nothing to do with guns. They conferred for a minute, laughed, and one said, "Okay, pacifist, let's see what you can do," and he handed me the pistol he held. There was a beer can on the ground about 20 yards away. I stood and slumped and scissored and hit the can six times. No one said a word. Not one of them ever kidded me about hating guns and killing again. At first, I did not like my own reaction because I felt good about breaking their stereotype of people who refuse guns. But then I realized that was not it. I had not broken their stereotype, I had earned their respect. Because I could "handle a gun"—kill—if I wanted to. The ability is enough. Colonel Wesson had said: "No matter what anyone ever tells you about handguns—target shooting, self-defense, quick draw—*never* believe them. The one reason anyone ever picks up a handgun is this: To kill."

Now, looking back, I think another thing about D. B. Wesson, and I am puzzled and wish I could know the truth. I knew nothing about his life before or after the lessons. In fact, I knew nothing about it during the lessons either. He never spoke of anything but his method. But there was something else in that look of eye, and the voice, and in the almost mocking nature of the method. I think the nature of the careful crafting of the method, the almost meditative process of it, made killing anything almost impossible.

Those who really knew him may think as they wish, but it was there in the stance, in the slump, in the scissors. I have done it, and this is how it hits me now. When I see a TV policeman take that combat crouch, frozen with arms out and bringing the piece down and gripping with both hands, I hear it: "Slump, damn it, slump!" And I know that we do indeed mean to kill.

Even I, using my own remembered moments as samples of an attitude, a way of noticing, wondered where all these shards of perception would fit. My gifting birds were not so kind as to leave me labels. How does my enigma of Colonel Wesson—for I may have made it up—inform me in any way like that of a young girl in Mexico, a city, a country store, the bleakness of a jetliner? Perhaps, I thought, the very fact that they gathered themselves together without my knowing why meant I had a search to do, not an unlikely collection to break up.

Naturally any sense of place will include people, if only oneself in a wilderness, for we carry our confreres with us. As my friend Ed Sherman says of even the smallest perception, "It's all in there," pointing to his head as Everyman's head. We are, after all, once-born human beings, fascinated by the fields of life around the others. So it is not just the humanness or the people and what people create which brought my group here together. I can only explain it by saying that each instance caught in mind, in what one might call a sprung perception. Every time, in an animal instant, I attended to small events about people and what they make. And that attention had an intensity which was unusual, more than the passing notice, though perhaps all our perceptions are still "in here."

What evoked that intensity may have much to do with the tensions of contrast, which is another way to speak of change. We have visions of such people as Amelia in El Salon Mexico, but we really know we shall never find her again—or even once. Her brief reality tripped my attention, for we are searchers and watchers.

And we are acceptors. With the onetime pleasures of travel in our shared histories, we still climb aboard the jetliner without noticing our diminished capacity to comfort our senses among lost sources which have disappeared so slowly we do not notice. And even in cities of the senses, what we find is the best of *Homo faber*, man the maker, and the insufferable worst in an environment which may lead us to join so many others of the animal phylum in extinction.

Other kinds of contrast were hailing me in the San Gregorio Store and in the puzzle of Colonel Wesson. The tension of these contrasts is the one most covert, the most overt, the relentless one between our lives and deaths. That most profound change we find after the last expiration is the one we understand least and yet cannot alter.

So much in the human niche looks out upon change, names it, predicts it, controls it, numbers it. But for the human animal, there is still one free place for the feel of change, for the gambol or the immovable stillness of our senses. And yet we often refer to our esthetic experiences as the frozen moment, as that minute point at which the artist "catches" a scene, a melody, a word, a mood, and our attention at the same time. And I wonder: Is there an esthetics of place? If there is, it lies in the unfrozen moment.

V

UNFROZEN MOMENTS:
AN ESTHETICS OF PLACE

The official arrangers of traditional knowledge categories, in this case philosophy, have placed esthetics under the broad head of "axiology," which is the study of theories of value. Along with esthetics under that heading are such things as ethics and theology. These are things in the area of selection, of choice, of what we want and value, as opposed to the "absolute truth" under other headings like metaphysics and cosmology and logic.

For centuries, esthetics was considered to be a study of the nature of beauty, and usually confined itself to the works of man, the creation of and response to what we call art. Sunsets, for example, did not qualify. Philosophers, critics, and various artists turning to *belles lettres* and definition wrote books which might have all been called, and some were, What Is Art? In the last century, more or less, the trend has been to include not only the beautiful, but also works of art which reflect anguish, horror, ugliness, cool neutrality, and even lucky palette accidents. Paintings of pretty flowers would not do as the sole realm of art, and there were those who could not bring themselves to call a Munch or a Kollwitz "beautiful."

Why estheticians did not concern themselves with things exciting to our senses but not man-made is not exactly clear. The non-philosophers, if there are such people among the population, went on thinking that murmuring brooks and birdsong were beautiful, that the deadly gyre of hawks and the lion at dinner were powerful esthetic moments. Perhaps earlier, more formal students of esthetics assumed the widespread and unquestioned position which separates man from nature, or, more dramatically, man from Nature, and history from natural history. It is one of man's common and self-limiting mistakes. The esthetic experience is the most intensely close response we have to any reality which is to include man.

An esthetics of place would have to involve sensory responses to everything present in any given place, which would include all five— or more?—senses, as well as time and movement. Nothing strictly human or strictly non-human. It may be a matter of what is con-sensed, a feeling of fitness between everything which is I and not I at any animal instant. The experience is not static, not a frozen moment, but a living one.

The canons of any art allow us to form emotional and intellectual meaning out of sensory responses. If there are canons to this art, this esthetics of place, they may be seen more clearly after some explorations in and out of some specific experiences.

12. The Waterfall Keeper

Watching and listening at the base of the waterfall by the plunge pool, I was looking for poetry in my head. A plunge pool is a fine place for finding poetry, but none came. There was litter all around, cluttering my solitude. Instead of poetry, images came. I was the keeper of the falls. Someone was coming from the stream from above. He paused at the head of the cataract, as if about to jump. Who goes there? I said. If you have not weeded smooth stones for passage on the stream, you must pay to enter. He leaped from overhead, and came up kneeling in the pool. There is no kneeling in the plunge pool, I said. He rose and backed away, under the falls, where I thought I could see him both dimly and clearly. Then he was gone.

I laughed to myself, sitting at Sacred Falls in Hawaii, telling my phantom there was no kneeling in the plunge pool. There were cans and papers and pieces of plastic on the trail and around the falls. In no sense was the place sacred. The official waterfall keeper did nothing to help. He sat far away in the parking lot to take the fee for seeing the falls. There were no cars there, though; the lot was empty, the attendant lolling along in a chair. We did not speak as I passed by on the road. I walked home through the canefield, feeling the toll taker knew nothing about waterfall keeping.

After all, waterfalls are special places. It does not matter how high they are, or how wide. Each has a world of its own: the stream, the fall, the plunge pool, the downstream side, with definite kinds of lives in them. I like small falls best. With them, you can sense what makes waterfalls such excellent places. Especially water itself, that primitive mover and shaper from which we come. It seems that when I look at a rill, pond, river, ocean, I strain to remember. At a falls, it seems clear, that sense of connection to water. It is not memory, for that is submerged in the passing of the genes beyond memory. It is not simply need, either, for that is clearly enough the message of thirst. There seems to be something known, persistent over millions of years, about the essential fluid. It is knowledge lost deep in the inarticulate brain—but still there. Every philosophical system makes assumptions essential to it. Otherwise, it fails—even if it seeks to be non-systematic. Water is a biological assumption without argument. The cells dimly sensed to most of us die without it. On the sensed level, we embrace a watery need without the formality of semi-permeable membranes. No one has to tell us about water, but water-falls are not well understood.

The waterfall abounds with sound. Large falls give off a steady, diffuse roar. A small falls trickles, it gurgles, it shooshes, it plays keys on currents. Bending low in a plunge pool, you hear the very sound of the spray. And see the movement. The spray hangs in the air as if it were always the same collection of minute rainbows. Behind it the cascade, above it the stream, below it the flow and spume of water moving on, back in its channel, the free fall over.

It is being able to attend to the whole ambience of small waterfalls which makes them my favorites. At large falls there is just the right sight and sound of water falling. The falls at Yosemite, for instance, are too grand in all ways for a waterfall keeper. Besides, a waterfall keeper does not care for that kind of audience and so many photographers looking over his shoulder. People from time to time are welcome, but the keeper must tend, not expose, his waterfall. Butano Falls, near Pescadero, California, has that tended dimension. The falls, small and lovely, is tucked finely away on private property, kept by those who see it, who seek its bubble and succor. The stream before it is small enough to watch, to follow the life before the plunge. And at the cataract, the falling flow can be touched and felt on the whole body. Once in a while a soggy branch, long in the stream, whirls off, disappears, perks out, dawdles about, and is off again, out of the silent eddies. Free passages.

In Cuernavaca, south of Mexico City, there are deep, steep ravines cutting through the town. Bridges cross some of them. Below the bridges and all along the ravines, the poorest people of Cuernavaca make their homes. I had often heard of a waterfall in one of the ravines, and I decided to find it. Down a muddy trail I went, passing huts and shacks and curious people, down to the base of the falls.

It was about 75 feet high. The feeder stream was narrow. Perhaps it had a name, but no one could tell me. At the lip of the falls, the ravine widened. Hitting at the bottom in a broad sheet of water, the falls seemed thrown, not just falling. The water came in long globs, splatted below, strewing water on both banks. I stood and wondered at it, decided to stay.

But there was no place to sit. The bank was almost solid with human wastes of all kinds. Paper; partly burned, charred, wet old fires; the universal plastic; pieces of old clothes and shoes; feces; cans of all ages; a decomposing chicken. People had said to me that no one went down there. They meant no one from above, from the walled, bougainvillea-bright houses. Clearly, lots of people went there. It was fouled, but then it was not a nest for those who came. I turned, and a few feet away stood a man looking at me with restrained curiosity. I bid him good day and he answered with a smile, pleased that this odd, wandering gringo spoke Spanish. I asked him about the waterfall and the debris around it. His shoulders went up and down and then as we talked he said there was no one to clean up, no one to name it, and that it belonged to no one anyway. As I moved up the trail later, I thought that a waterfall keeper would not have helped anyway, so I did not apply, but just plodded back up the trail.

Years later, I was still looking now and then for just the right waterfall. Certainly a place called Seven Falls was an immediate attraction when I heard of it in Tucson. The falls were near there, so, with canteen, lunch, and book, I walked that way one morning. It was cool and clear, the trail gentle, lined with birdsong. I passed a strange saguaro cactus with one mutant arm shaped almost perfectly for the capital of an Ionic column.

At the bottom fall of the seven, I put down the rucksack and looked over the plunge pool, peering down into a sky blue. I turned away and climbed up to the second, smaller falls, and then the third. But the first one had stayed in my mind, so I turned back.

The stream was small, the waterfall not over six feet high, the plunge pool about 20 feet across. There was something perfect about it. The proportions, the air about it, the melodies of sound?

I sat on the bank near the spillway, began idly to look around, then look closer and closer. Finally, a movement along the lip of the spillway started to be odd to my eye. What appeared to be a dark coating of some mineral deposit was moving. Looking closely, I touched it and discovered a mass of tiny worms about a quarter-inch long, attached at the base, taking nourishment from the water passing by. What a fine habitat they had, but the water would stop, was seasonal.

As I watched them, flicks of orange just above the pool surprised me. It was half a dozen bright orange dragonflies, flitting above the pool. One flew nearest the water. Another above it followed every move of the lowest one. The rest tried to get between them, but the one following closest darted out and back on brief sorties to keep them away. On one quick flight out, another dashed in.

The one nearest the water was a female, the rest males. They flew patterns over and over, until one male was able to hold off all the rest for several minutes. As the female slowed and began to hover, the persistent male went down too, and the rest left the chase. The male and female, tails bent, joined briefly, then the male rose just above her and flew in circles. The female flew around slowly for several minutes, then went down and held barely above the water. With quick dips of her tail, she deposited the eggs. Then both were gone.

I had just seen incredible things, by merely sitting at a plunge pool.

It is a personal thing, of course, but for me it is inescapable. If I watch even worms and dragonflies long enough, I feel related to them. Which is certainly true. I neither know nor care how far or close we people, worms, and dragonflies are on the scale we have given ourselves to measure by. I feel closer to people, being one of them, and closer to dogs and trees; but, after all, ranking and weighing and comparing is one of man's characteristics. Worms and dragonflies have other qualities, neither better nor worse, but different. There it is. I find my animality neither gross nor sublime, but rather a key to marvels like small waterfalls. And my attitude about waterfalls is neither true nor false, romantic nor realistic to me. I wish to convey nothing by it but a simple, personal joy, which I can both speak of and never utter.

If it is true that a writer's finest moment is the one in which he decides to remain silent from then on, I know what I shall do when the time comes. Given those falls still there in Arizona, I shall return to that lower plunge pool and apply to the worms and dragonflies and people around as waterfall keeper.

13. Adirondack Spring

In our flanks and feet and in the dancing parts of our heads, we always remember certain Springs when we grew up. Forgotten Springs are lost and gone and have no rhythm. We may let them go, attend to newer lives without season; or we lose a fragment, or keep a piece here, all jumbled in memory. A big dance or concert simply vanishes, but a daft little day of cider and ice skates just will not leave. How we learned what "sine" means, for example, may remain a wet and murky canvas, but certain faces are eternal, without any effort.

Spring jostles. Its hints have their keening nudges, send signals like drips of snow. Fireplaces act differently in the air, not making autumn. Ice and snow sink smoking into the earth. About Easter, the Spring doldrums set in; teenagers get the surlies. Everyone wants to move around. Just that feel to move.

We must always know that, even earlier than Spring. Some mute sense, conceived in winter. In the high school years, there were parties to come, and track meets, and the whole revival of flesh behind hockey-puck bruises. That is never quite all, though.

It was vague; just that forecast of moving around. Not that it portended much that was known; nothing like the joy of opening or the fear of leaving. Real things mingled. There were still parents to handle, and, for me, an older sister; and trips we would take at no certain time, though we moved with them as we spoke of them. Sixteen is good for just that. One wanders and wonders, perhaps smirks more than is good for the mind. My friend Dick and I may have wondered already what we could do in the Easter vacation, but if so, we kept falling into a bunch of nothing. So what happened first came on its evolutionary course from an unlikely source, my stepfather.

Never in the years I knew him was I prepared for what my stepfather might say. In the combination of wisdom, intelligence, imagination, and sensitivity, I have not met his equal. Naturally I did not know that until later, some years after he died and I found out I loved him, as his own children did. But then, in that Spring, I knew I liked him a lot. There are times growing up when nothing can get your attention, but I listened to him, always, fascinated and puzzled. He created things like this with every word he said: curiosity, suspense, rhythms, teases, laughter, sadness, a world he knew.

His world was so huge and mysterious to me that I listened with the awe and the thirst of any ignorant being who may want to hide but can't help taking a look. He longed for me to know the world, to know how to learn. Good teachers always want that, and the others have jobs, not a way of life, and turn out products of the school system. Which is appropriate for a society in which grown men in various costumes chasing around a field make more than teachers will ever dream of for evoking agile minds.

My stepfather did not know anything about that. He knew about the endless journey from leaving school in the seventh grade and sweeping the floors of the Railway Express office in Brooklyn to becoming a vice-president of that company. Such things happened. Such people happened. Yet that was not the rare thing about him. He had an unrepentant curiosity and an insatiable, endless ability to relate what he had lived and read, poetry to history to wildlife to economics. He was a self-taught man who had also learned how to understand.

Even about kids and dogs. He recited poetry to my dog—very wisely, it happened when I was there to overhear—and swore the dog had a favorite line. It was from Milton's *On the Late Massacre In Piedmont*, "Avenge, O Lord, thy slaughtered saints, whose bones/Lie scattered on the Alpine mountains cold"; and the dog would perk up at "bones," hearing the affection in Poppa's voice. Without knowing it, I attended as much to him as I did to the luxury of my age, which allows us not to care about not knowing anything. So the dog and I did not know how much we really learned.

So it was that Poppa taught, casual, relentless, the full measure. He would never slight the esthetics, the empiricism, the experience of event or place or person, or the art of each. When he heard our restless voices speak of Spring, he began to tell Dick and me about the Adirondacks. We listened to stories of places and people in those far-off mountains where the primeval woodsman learned more than we could in the pockets of woods in Scarsdale. The Adirondacks. The North Woods. This was serious. We had had vague ideas of sneaking girls and a boat and fishing stuff into Lake Byram in Armonk, just a whiff away from home. No more. Our spheres had leaped north. We listened. We heard new things become real things. Books had voices, maps had faces. And Poppa knew the route.

Poppa had a plan. It was our Spring, but he would lead us to it. With us, he would take the train he had taken so many times forty years earlier. He would see his towns, and there might be people left. Dick and I were not so ignorant that we failed to see his trip as well as ours. He was treating and teaching us and himself. The last lesson came to this: At Speculator, he hired a car (one still did that) to take us to the road we were to walk to the Miami River. We stepped down from the car, he told us where we were to head. Perhaps from years with the railroads, he took out his watch, looked at it. He smiled, said, "Okay. You're on your own," and got back in the car. We waved him down the road, and we were on our own. In the Adirondacks.

Of food, we had more than enough for the five days, and both of us were experienced campers, could cook all those joyfully horrible things young campers cook. We had the main gear: rod, reel, flies, creels. Ready. For anything.

The first night we found a good spot to camp, near the road, and near "the river," which was a disappointing trickle where we found it but was supposed to be a fine stream up a little. So we made camp, ate, put on the mummy bags. Our food-laden packs were hoisted in tree branches. In those days it was not only permissible, but a sign of knowing the lore to make a layer of tree boughs under your tent area, which was outlined by logs or heavy branches, and trenched to keep possible rain out of your sleeping snoot. One did not want a cozy tent with the water coming in under it. So we did all that. We stashed the rest of dessert—peanut butter, crackers, and prunes—in the tent. Then we slept.

We were not ready for anything. Easter vacation. One thing we were not ready for was snow. It was a foot deep against the tent when we tried to flap it open in the morning. Just a pup tent. I think at that point we were so stunned we never really got our sense of reality back, though snow fell whitely in our faces as we got out, not very dressed for the day which fell so quietly on us. The world just does not do that. We reteated into the tent, dressed as snowproof as we could, and we spoke almost not at all. We could not wash or make a fire to cook on. Cold food we could eat. Can of corned beef hash, beans. No one carried self-contained stoves then; one made cookfires with smallwood. The snow was very heavy. We got back to the tent. Something had to be done. First and unconsciously was to take in what was happening out there. What did it mean to us, what could we do, what should we do? We had a conference, lying shoulder to shoulder, both looking up into the olive drab canvas which we could no longer touch if there were to be no leaks.

The choices came down to staying or going, then to staying. Surely the snow could not keep falling, would melt quickly, and we would fish the Miami as we had come to do. How could we go home just because of a little snow? Were we not there for what we found, for what happened, for the great challenge of the Adirondacks? We would wait.

Surely the snow would not keep falling. That day we honed our gear, we ate cold, we crawled in the tent. A round of Twenty Questions, of Ghost, of sleeping in the closeness of the tent. It had come a little in on us, for the snow was not surely stopping.

In the early morning, I tried to push the flap aside, and it would not go. I spread the gap between buttons and saw snow. I spread a higher gap, at the top of the flap. Air. The tent was almost buried in snow. I unbuttoned the unpegged flap, pushed the snow aside. Outside was cold and damp and heavy and silent. It was far away. I had seen snow before, close and distant. But I picked a loose bunch and spread it out in my hand. Flakes, of course, going water there in my palm.

Buried by snow outside, I was intruded by snow inside. And that was only the start. We dressed and then fought forth into the snow for the packs we had treed so long ago. Stew, we had, and cold; with beans in that curious, congealed crust of cold bean juice. There in the falling snow, falling into the stew, into the beans, into our heads. We did not know it, but there was really no escape. By noon the tent was covered, and even in its comfort, the snow resided in our warm bones. Wait, we decided. Leftover lunch. Prunes, peanut butter, and crackers for dinner, more games all the way into sleep, into the fitful neighborhood of falling white dreams.

There was, oddly enough, a morning. The tent was almost touching our heads. We dug out, up, had our delicate morning toilet with white snow dug to wash and yellow holes of relief in the snow. Then crackers and peanut butter, over hard, and no prunes were left. The snow fell, it heaped like heavy dancers keeping distance as they could. The deep bottom packed; the high, new snow following down made us look at each other. We would have to walk out today. Or perhaps never.

We decided. Dress for the snowy hike 22 miles to town in knee- to hip-deep flakes. Leave everything behind. We could find someone to send it when the road was clear. Sixty pounds of pack apiece would not be a good guarantee to make the hike ahead. We made everything as waterproof as possible and waded to the road.

You expect bells at first. Or perhaps not everyone does. I did. Not the cornball bells of sleighs and giggles and cheer. Distant bells, not too large, tolling vaguely but so clearly the sound rolls on snow and comes pure to your ear. The silence held fright, portended deadness. We livened as the anechoic crunch of our feet came back to face us. We moved. The road wound, and we stayed with it. For a hundred yards we said nothing. And then stopped. The snow was above our knees. It was not possible. We spoke with awe of space to go and time to move. There was no choice to make. We had to move, to plod. One foot, one foot.

My feet kept going and I looked and listened. Snow lined branches, nude ones and green ones had popped back after the heft had fallen. Lumps and holes gathered at snow level, making tracks like amorphous birds leaving the scene. Out away, ahead, the trees were deep and did not grow right. They seemed tended, and stuck there without roots, without form to be in the earth. Snow is like rain, of course. It is only water. But this snow was implacable, was undeni- able, was immovable, was unbeatable, was deadly. And it was the most beautiful thing I had ever seen.

To my young brain that did not make sense. It did not even seep in around the corners for the first few hours. There at my feet-watching eyes, off in the silence of dark trees was an incredible thing to see: snow. What had I missed before? Perhaps only being forced to bear it, go through it, to be thrown with it and denied choice. When the air thickened with snow, I stopped.

"Hey, Dick. This is beautiful."

He looked at me with the immobile cold of a frozen face. It cracked a grin.

"You're right, C.J., you're right."

And we threw some snowballs and ate some snow and laughed for the first time and walked on, one foot, one foot.

In the afternoon we passed a cabin by a whited lumber camp, defunct for the winter and, now, into Easter. There was someone there, and he was pleased to have us. Caretaker. Clearly we were fools to be out there, but our youth and his loneliness excused us. We warmed and he offered us hot food: bacon, pancakes, biscuits. We thought him a fine old man. He did not even keep us for stories. Probably because he was looking forward to the $2.50 each he charged us for the meal on our way out. Somehow it seemed to rankle against the code of the wild-storm-lost-boys routine. We hiked on.

The afternoon grayed down, and we kept going in our woodcut of black trees and pure snow. Scene after scene moved by, unfrozen moments on an endless, forever moving tour of nowhere. I would stop to listen to no sound, to look at no movement. Yet nothing was still, captured. I moved on. It may have been the slow time of exhausted walking that made me attend so. And then it was dark. The snow kept a shine, the sky a blurry lightness. My private moments grew dim.

It was eleven when we reached Speculator and got a room. We soaked our feet in cool water, ate our last chocolate bars, and dropped to bed. I recall thinking I would never see snow in the same way again, that there was no longer such a thing as just snow.

14. The Islands

There are times when a city offers nothing but its open face. That is especially so for a city well known as being ideal. In that most glorious of cases, a tour becomes not just a sightseeing of history lessons, but a present of stunning facades, even if they are false fronts without questions or answers, nothing behind them. No one would think of going to the welfare files to get to know Tahiti better. Or even Stockholm, for that matter. In such places, we want to see smiles, the ideal, the face unrevealed by its eyes.

That is how Honolulu first came to me, its sunlight bearing a green that shone like gold under the stringing clouds of the valley. It was not unusual then to see the islands first from the crowded railings of a troopship returning from Korea. For me it was unusual; more, it was unique. Never had I seen such greens, along with the shores and houses shimmering together like the sea itself. Nothing had been on that sea, not another ship, for twelve days, and the verdant mass of Oahu rose from the moving waters like an enormous Atlantis ready to sink and leave us alone again. Then we anchored, sharing our fate with the island.

Already on that day, with the first look, I was condemned to almost ten years of eternal return to Hawaii. It could be more. I don't know yet, though now it seems I can either go or not, that I have finally given myself a choice. That is everyone's Hawaiian problem: to go or not, and where? Perhaps it is the islands which have given me a choice. I kept wanting to go back, and then going back, seeing, hearing, smelling, feeling the changes. The more I went back the more there was to sense and the less I wanted to sense. You could hardly avoid the yearly stack of new hotels, the traffic, the swirling people, the polluted skies.

When I first arrived, Hawaii was the kind of place where there were always hidden secrets, those forever elusive essences which stay behind when one says farewell. They are of a lost past or a hoped-for future or an obscure present, one in which the frieze around a city is its only decor. You want more than outlines, and that is why, as with a human face, the language of the map will not suffice.

Anyone would be willing to write about the sun, the beaches, the happy people of Hawaii and their lush surroundings. In my first days there, as now, too many words spoke of the place as a paradise, which was as unjust as it was common. Hawaii shares with every other place on earth the impossibility of being "got right" by word or picture. Only the place, any place, can graft its mood to you, and you begin to grow it yourself. Hawaii never lets you alone. It is so often assumed to have been captured, when in fact less of its depth has been felt and more of its surfaces transported than perhaps any other land so rich in sun. Travel writers tend to count landmarks, give histories, pose facilities, and trace the people of the isles. It is a diverting, even informative tour. But one can pause to savor what the eye is given.

When I first arrived, that pause was easy because so much was given. The conflict of town and tank, eternal in wars from ancient times, is nowhere more apparent than in Hawaii. Socially, the private among the ranks makes his way sub rosa. Out on a walk in the wild, he, like every citizen, finds the olive-drab, officially labeled water faucet sticking up in the jungle, looking like taxes due, a reliction of the Pearl Harbor mentality.

I simply did not enter the conflict. Quickly, I went to the other side of the island, the country part that still looked like the fable I created from the rail of the ship. And there it was. I found a spot at Pat's At Punaluu when it was still tiny, awash in the South Seas, and sagging under jungle rains. It was run by Pat and Iris Hallaran, and they later called their area "the Irish Coast." Both became friends and mentors, their place my home on weekends.

To earn my room, a cleared spot in an old bungalow, I was a busboy and stocked the bar. For board, I discovered a world that was not only an alien land, but became my den, my escape, my habitat every Saturday. The ocean there was broad and open and gentle. You could sit and listen, feel the breezes, smell the tides. Far away on both sides of the beach from where I stood, the trees and the sea shimmered in the eye of light. There was a sweeping to the air, to the sea. There was a long, high, warm cleansing all around. I thought at first I went down there to go in and spear enough fish for a weekend of meals, that I waded in and kicked off into those waters for the undersea hunt. I did not. I went in because I loved to see it and hear it, to be alone with it, the gathered life of the sea. It was a lesson I had learned and would learn again. I did not wade in streams for fish alone, or climb mountains for thrills, but simply to follow the tuning of my own senses. It was just the way in which, later, I was to listen to Bach's music or stand with stupified eyes before Vincent van Gogh's paintings.

Even before I got into the water every weekend, it started on my way over as I left a week of army behind, sitting on my motorcycle. I could smell the very pineapples, feel the scorch of the sun, and yell delighted defiance at the rains as I passed through them. When I got to Pat's and into the snorkel, picked up the spear and pushed out, the world of air was gone. The sound had a ting to it, the smell a salt about it. I had been told what to look for, what was good for food, but there were times underwater when I forgot I was even there for food. All I could do was look at the fish, the formations of the sea floor, the eddies of sand as a fish snapped around a stony head of coral.

Stopping to listen, I could hear my own breathing. The sound came from within my head, went out into the waters and back, as if someone out there were breathing at me. And always in the background was the constant, mysterious, hollow ring of underwater. Had I been a hunter up in the air at those moments, the sounds of birds would have made me lose interest in deer. The reality of eating, though, always brought me an aim that worked often enough. No trophies. No prowess. Food.

Coming out of the water, sometimes after eight hours there with pauses on heads of coral, I was starved, ready to cook fish, give some away. Always after a day like that there was a drying and warming, a glow of return, the twinges of muscles well spent. There was also a curious energy, a joy that I had come with food from the sea. Perhaps no one noticed that I walked around happily in wrinkled skin, warming in the sun. I felt, but probably never shared in words, the abundance of the sun and sea and friends, though I thought I did. Years later, I saw us all—people, sun, and sea—differently.

Every land of sun knows a certain plenitude. In Hawaii in those days, there was a way of life which went beyond what one is inclined to call "friendly people." It was a kind of level exhilaration at being there, a feral knowledge that goes past what is simply given. Land, sun, and sea were taken into the lives of the people even more than the quick smile to the stranger would seem to speak of open hearts. The unity of an island people was here translated by the land, by a love and closeness of its fruits and a share in its bounty of the sea.

If this unity was unspoken, if it was never idealized by its own people, it was still why greetings among strangers were not exceptions. And it is why the pagan mixed with the orthodox. The land, with the sea, was alive in the hearts of the people. No one said this. It was shared without being spoken of, as the rush of New York would be by its people were they not so self-conscious. There was no rush here, but that is not to say the islanders were languid—no more than their share. It is only that this closeness was bound to the land and sea in a tempo of natures which kept to its roots, grew like its leaves.

The quick smile was a reaction, but the people were not easy to know. Those who shared the beaches, the mountains, the sea, the special language of Hawaii, and the nurture of its soil did not escape a pride in their home. To know them, one had not only to know their customs, but to cherish them. Not because of the people, but because of the customs. If you did not like country luaus under a shed where raw fish was served, it did not matter what you thought of the guests, even if you loved them all.

In town, whole families descended to the beaches after work. It was part of a daily ritual, a true homecoming. The spirit freed itself from all that was a task. There was no saving up for a holiday. Surroundings were not separate. The sun was in the open houses, on the beaches, part of the people, a way of looking.

If there is an esthetics of place, its hungers could be indulged, overindulged, jaded, sated, glutted in Hawaii. Through your eyes and ears and nose and taste pass, one after the other: flowers blooming all year, an impossible variety of greens, superbly steep and eroded mountains, sounds of surf exploding on the shore and whispering back into themselves, the manic flavors of guava and papaya, avocado and banana, the soft winds bringing hints of damp jungle earth and the primeval smell of the sea.

On the surface of that sea were those who lived by it, for it. They were either in it or on it day by day and in the night, looking for fish, squid, anything offered in the shallows or deeps. Theirs was a work of harvest, of spending energy and good times for the feast of a family or a boat load for a fishmonger. But even the men who made their living at it never invested their souls in a distant tomorrow. Not that they were imprudent, casually smiling natives. They merely knew their work, their lives, what they needed, and what they gave.

This was no fantasyland, though. There were those with ships and tallies and records of work and goods and where it should all go. There was today's pleasure and there was commerce. Ships and planes brought to the agents all that was needed by a modern people. There was an old and solid and diverse business—production, advertising, agriculture, banking—all in many tongues. There were the old families, sons and grandsons of the missionaries, and there were the new young men of the trading world. The bright young businessman suited up for his task was far more common than beach boys and exotic maidens emerging from pools at lush waterfalls.

An awareness of business was nowhere more striking than in the bustle of tourism. Its active, uncoordinated shotgun approach was impossible to appraise then. There are times when hard cash is so frantic that it has no mystique. In a sense, the rush for tourists was a frenzy, a kind which led to great care about words and pictures but none for reality. Shown without shame was the fabled Hawaii, the Hawaii which called from the heart of the noble savage, the Hawaii of a romance with paradise, the image from the primitive South Seas. And then visitors would arrive to the surprise of paved streets, stolen cameras, and chain store clerks telling them there were no bathing suits in stock because it was January.

Everyone knows that no place on earth is paradise. Hawaii was no exception, even through the people were aware that the sun, the sea, and the land—the neighbor—were all in a conspiracy to provide heaven. All was changed, made, by that. There were extremes. One could bask in the sun or rush around on errands of entertainment or even work. The extremes mingled and they clashed, but the people who lived in the Hawaiian sun accepted all it offered, from ease to anxiety.

And indeed there were places of misery, outbursts of rage, cities

which like other cities had their hot, despicable slums at dead end.
Here they only seemed worse—larger, dirtier, and incredibly desperate. One is tempted to say that this extreme was only seen as worse
because of the other—blue skies, waving palms, friendly people. Yet
it was not the contrast which made this poverty squeeze at the mind
and stagger both resident and visitor. It was that a land allowed one
excess must accept others. That is a realization the brain must live
with for a while. Here there are excesses of darkness in the midst of
excesses of sunlight. Out in the sun there was nothing to suggest
hatred among the various peoples of the islands. But on the run-down
edges of the city, hostility was a way of life. The last "ethnic group," as
the sociologists said, to arrive for work in the canefield is the one at
the bottom of the social scale. Perhaps. It just happens that for all but
an arrogant tabulation, every group in this part of town was the last to
arrive. The pool halls, the fortune-telling stalls, the streets were full of
last arrivals. Such people as these in a sun- and sea-minded metro-
polis communicate with loud, thin laughter or with knives. The
prisons held a number of those who spontaneously committed all the
crimes of too much sun. But the prisons tried, and the state did not
kill.

     Perhaps that was because revenge is a feeling not so acute under
skies filled with sun and the foam of a warm sea. Or perhaps there is
no need for eternity's gifts on a soil which abounds in its own. Seasons
are not marked by the bloom of a flower or its withering, but by *which*
blossoms fill the streets with aroma. But missionaries from other
lands have never forgotten eternity. There are always missionaries in
Hawaii—or in any other clime where the senses are offered so much
pleasure. In Hawaii, the missionaries from the past are represented
by the present wealth and social position of their heirs, as well as by
the many churches. And the missionaries of the present are most
conspicuous in the form of well dressed young men distributing the
Book of Mormon. There are many other churches attempting what in
Hawaii seems to be a great austerity program of the soul. The
churches have their converts, but the deepest religion seemed to be a
communion between the people and their land. If God was dead, it
was less interesting in a religious sense than it was for what it told us
of our feel for the land. Behind that idea is a more basic one. The earth
is dead. We no longer feel connected to it, whether we are farmers or

taxi drivers. The earth is no longer the giver and taker to which we were supplicants through our gods. We may survive with impunity if the gods disappear, but we cannot live without the earth.

But the earth was still here. If the abundance of beauty merely jaded the taste of the visitor, it quickened the spirit of those who grew in that abundance. Everything was accepted. Life and death were met without pretense. It seemed to be an impossible simplicity. Yet if true goodness lies in never having to ask what goodness is, true worship lies in never having to ask what its object is. Such a closeness is lived. This kind of life with the land is like a daily vestment, donned wearily or with omens grinning, in a city or with feet on the earth. It is a certain profound paganism which comes before any formality of religion and disappears after it.

Clearly this was a land of contradiction, as one always says of any land where the people are offered and accept the harshness of extremes. But as it suffered for what was denied, it bathed in the bounty of what was given, and it chose the bounty. The poor on welfare, living in the ragged quonsets of the countryside or in the jumbled slums of the city still went to the beach, still knew the sun was there.

If the needs were many, the poverty great, the commercialism disappointing, Hawaii was yet bound together by a unique gathering of values. It had a feeling of what the senses shared and found nowhere else, for here was enough.

Of those earlier years spent in Hawaii, my memory warns me that perhaps what I recall is a place that never existed. That a place or a person never was real is a chance one takes with words. Yet, like the man who wakes up one day to find he is an athiest, I want my faith back, even if I know better.

For that early Hawaii no longer exists, if it ever did. It is said there are reasons for that. Too much tourism; land monopoly; the high cost of living in general and island living in particular; that state graft and greed are larger than their past counterparts in a territory. These may be problems, but somehow they do not speak to me of what has been lost in a land which once lived in a state of esthetic grace.

What is common to these problems and causes, these losses? An attitude, perhaps, a change of spirit which no longer commits itself to land and sea. Not to the sea? There are still those who are in it, on it, under it, but no longer do the people of Hawaii live with the sea. They look at it, some go to it, but the mere gurgle of it and the bounty of it in their lives is gone to most of the families there. Fish in the market is as high as shipped-in meat, the beach is too much time away. The land promises more.

Land, land and its new kind of bounty has replaced the sea. But no longer is the land for pineapple, cane, fruits which spring from the soil. Back toward the mountains and out on the beaches, the land has a different voice. It calls no more for growing things, but for use. There is no sea, there is no land, there is no sky.

There is space. Space is for sale. Space is used, filled, paved, built on and called "open" by developers of golf courses. It is never kept. Land, sea and sky do not belong to a modern society, we are told. The Hawaiian land was green; it will become neat lawns and macadam. The sea was blue, green; it will become a sluggish brown oil. The sky was blue and brought clear waters; it will become a gray slit between buildings.

Why has Hawaii let its sun, its land, its sea become space, come up for sale? It is not simple greed. You understand that when you realize that its languages, its history, its archeology, its anthropology, its literature, its art are of little interest to its people. No one knows why the people have become self-conscious without a conviction to be themselves, define their place. The main pursuit seems to be imitating the so-called affluence of the Mainland.

Before statehood, one of the voices heard in the Hawaiian land said that islanders ought to become "first-class citizens." Perhaps the people of Hawaii have become first-class citizens. But like their brothers in Florida and California, they may one day be puzzled and wounded by the fact that a people and its place, like a person, can look with deep sadness on its own fulfilled hopes.

15. An Old And Simple Mine

One elevator went down on one cable. At night, the twelve o'clock shift started in hubbub. The hopper, where the ore was collected overnight; the office and work rooms where the shift dressed; the elevator and its tower were all lighted by dim bare bulbs. Lamps on hard hats flicked around toward the cage to take us down. Since there would be no ore trucks until morning, the sounds were muffled, long ones: the whirr of the cage up and down, compressors somewhere, the bells signaling from underground to the elevator man and back. I could see the operator pulling and letting go a dazzling number of levers, as well as the bell. Men going up and down depended on his mind and hands, his quickness. This man was even said to be sober at all times, though in a year I never met him.

The night shift seemed easiest to me; at least you went down in the dark to the dark, and the pay was a little better. This mine, the Santa Rita near Tucson, Arizona, had been worked first by the Spaniards 200 years before. It was now deep workings and its levels were long rays underground. I was anxious to see it, since I had been working on the hopper emptying ore cars for two weeks waiting for hard-toed shoes in size thirteen. Why you could work the hopper without the shoes no one could tell me. Maybe because if you got your toes between the car and the rails it was on, it did not matter what you had on your feet. The ton of car and ore would flatten your toes like pennies on a railroad track under a fast freight.

When the shoes had arrived one night, I looked up and said goodbye to the hopper, checked out a hard hat with belt battery and lamp, and joined my shift on the way to the cage. Our destination was 2,000 feet down. I think I took that news like a veteran. The routinely fast drop of the cage froze me and I was convinced we would hit bottom, if there were one. The time of year was November, chilly at night above ground, chilly on the first part of the trip down. No one else said a word. Certainly I did not.

The asbestos pipes bringing air down to 2,000 feet pulsated slightly when I stepped off the elevator and saw them hanging from the mine timbers. The walls and ceiling were ochre, stained dark here and there with water; the underfoot was mine-car tracks in mud, the lights dimly yellow.

It was hot as a blast furnace, wet as the tropics, and smelled like the bottom floor of a dungeon. Everything was damp and dripping, the drops heard like the flow of a creek in spasms, the mud like clay. It was impossible to tell if it was sweat or just water that soaked you before you had walked fifty feet. The sky was far away; the night had just begun. I was mute.

To suspend my own disbelief, I lagged behind slightly and listened to mine timbers as we walked along the level. If you put your ear to a timber and can hear it creak and groan, you might be safe and come out alive. If the timber is silent, you are in danger because it has gone the limit of stress. Its lack of response tells you to get out. I put my ear quickly to each timber and heard the music of healthy wood. You learn things like that very quickly in mines, and use your knowledge even more at 2,000 feet.

No matter what you do at that depth, you and everything on you will be soaked by the time you take the elevator up. You do not cart your lunch in a paper bag. If you take cigarettes down, they and their lights must be in a metal or plastic container, waterproofed from the very air itself. You know your footing will slip a lot, your clothes be wet and dirty, your hands grimed and clammy. At your lunch break, from 4:00 to 4:30, you must eat all you have; there is no other time to take a bite.

Walking along with the crew in the wet and heat, we stopped at a raise. To connect levels in a mine, raises are blasted up or winzes blasted down. This raise was my position for that night and many more, along with a partner, Dal Radcliffe. Before we went up, I could see the lights of the face off in the distance of the level. The face is where ore is drilled, blasted, and taken out—the work of the mine where it counts, where the pros are. Having been hired as a mucker, I never saw a face up close. A mucker shovels waste only, not ore.

So. What is needed is a connection between one level and the next. The powder monkeys blast up or down. The muckers shovel it all away. All night long. In the dark and damp, stripped to the waist, you simply shovel and sweat. You do it at a pace you can keep all night long and yet satisfy the boss. On the night shift he is usually a senior or graduate student in mining or geological engineering, and he comes around to poke his head in at irregular times, checking your pace. You are not forgotten, even at 2,000 feet below and working on waste. Nothing can be overlooked, not even muckers. And if you muck long enough and well enough, you might become a miner, get to the face, send up the ores of copper, lead, zinc, and silver.

The sound of the shovel's bite does not go beyond the shovel. There is no echo. The only traveling sounds are long, back and forth in the tube of the level or up and down in the shaft. There is no free air, and what there is comes from the pipes, which seem to be on an economy aerostat. By 4:00, at lunch break, the air is fetid. The oldtimers talk, words cut off at their mouths. The new-come men in the mine take a while before they have energy enough to both eat and talk. In a week or so, the rhythm is sensed. Humidity and heat, work, truncated sounds, jargon, jokes, and the air strained through water are all accepted, as if taking everything as it comes makes the job easier. Sometimes it does. Here it did.

The trip up in the morning was no respecter of dawn as it is known. Light came first from the top, not the horizon. In winter especially, but summer too, it was always cold after the dark yellow tropics below. When the shift changes and men instead of ore cars are on the elevator, there is relief for everyone. The thin man up there operating the cage plays small tricks with the bell; he is reprieved too from nothing but ore cars all night. As the cage ascends, light whooshes in, you stop. Every time, if you faced out, the light morning hills were brown there. The light, the air, the rain never will fall down where you were. Man wires it and pumps it, and pipes it down for a living.

Almost a year after the first night under, the cage came to a stop and as usual I looked out toward the hills. Struck by the sight, it was hard to move, to stop staring. The desert hills were covered with a thin layer of snow, white for the brown my eyes were to meet. It is probably not strange at all that I decided it was spring.

The first canon of an esthetics of place is perhaps hidden by its own clarity, missed by its simplicity. The excitement of all our senses at the same time focuses on one event in a given place at a given moment which endures, lives, in time and memory. It is an animal instant which involves every sense keyed to an event. That event may be the joining of the senses on the notion of tending waterfalls, a feeling of the absoluteness of snow, the living continuity of a place which dazzles the way Hawaii does, or the suddenness of the odd snow on the desert after the strange sensory stimulation of the capsule of the mine.

A concert of the senses reminds us suddenly that there is a dynamics of the senses, a movement in our awareness of everything at once. Our senses always work together, not in bundles with this for ears, this for eyes; just as the nudges which move them come all together. It is in the moment all senses focus and we live with our attention that the esthetics of place puts us with ourselves and everything in our immediate place, our site.

Obviously, the senses we have of time and movement and other people are part of the focus of attention in an esthetic response to place. Those, along with a feeling of the earth itself and of other animals and the plants, are the subjects of the kinds of perceptions by which we discover what a sense of place is. If those subjects are the forms of an esthetics of place, analogous to the symphony, concerto, sonata, for example, then it is the senses themselves which compare to notes and chords and melodies. If our esthetic response to place is to be at its richest, we must learn about our own senses of sight and hearing and taste and touch and smell, to begin with. And we ought to learn them down to the physiology and anatomy of their organs and how all work together in man and other animals. Especially we should know how they work in the case of the individual, how I orchestrate my senses and mind and the rest of my body with what exists out there. One may speak of a "cultivated" taste. The best way to cultivate the attention we need for the esthetics of place is to tend the garden of the senses.

We do that very badly. In our schools we do not learn the esthetics of anything, not one of the arts. Literature is plot; music is performance; painting is poster graphics. About the richness of our senses and our feelings and the life around us we have merely biology and science, as if information made for us the best lives we could make.

It is possible to speak of richer minds, neither higher nor lower. The simple intensity of well educated senses can happen to anyone, anywhere, anytime. With a little attention, the human animal does that. There is no prescription, no "how to."

There are remembered perceptions, left at random by the gifting birds and picked up like pebbles, some or all from the same trove of nesting pieces. I relate them one to another and fit them here and there, as if they were as insecurely held as our own lives. Thoughts and attentions grow from them, piece by piece, never finished.

VI

THE EARTH INSIDE

To omit the literal meaning of the earth inside in the beginning would be a lapse of attention. That the earth is indeed inside us is so obvious it is everywhere ignored, the feeling of connection lost.

It is said that because our blood is the same in salinity as the sea, we have carted with us up on land a reminder of our source. That may be; but we do know that now, any given now, the ocean sustains and imposes the conditions of our land and life. For man the animal, the ocean occupies many niches, shaping the water cycle, wielding the power of winds and currents, witness to the maze of forces spawning continents.

And the very earth itself, the dirt and rock and sand and loam and bog and silt, is captured as well in our veins. Minerals, the vitamins and oxygen synthesized by plants, a selected diet from the periodic table, all course the chambers of the human animal's body. From the earth. We are precisely creatures of the earth, bone, blood, sinew, and nerve. And our hands are the umbilicus of the species. We *know* those things.

The other cord for this animal is the brain, working, which is merely another way of speaking of the mind. Our bond with the earth is an existential one, live and finite, real and palpable, working every second. We can *feel* it, though, or not. Living that feel of the earth is a matter of choice. Ignoring the fact of the bond would simply be a final mistake. Not sensing the bond has made us a poorer animal.

The outward signs of our ties are as easy to see as the gifting birds' tidbits, and perhaps as casually elided or ignored because to attend them makes fresh puzzles of where they fit. The earth has, for example, no boundaries. There are no countries. To feel that as well as know it is to understand that a man in Lima is not just another Peruvian, but a fellow creature, born like I am in a shared humanity. It matters, just as a certain attitude toward war does.

The earth is our home, the source of our nature and nurture. In war, we are not only killing people, but destroying our home, fouling our nest, the ultimate human and environmental disaster. At one time, war was waged for the profit of the winner. Then, as some degree of global cooperation became a necessity, the loser also profited because both winner and loser had to be brought back into the economy. But we have now reached the point at which no one will profit. There will not be enough people, tools, or toys left for an economy, for anyone's profit. Perhaps if there is no nuclear war, that will be one reason. We have reached an historic point in man's life at which there will be no return on an investment in killing. We have reached the limits of greed. And may still exceed them.

That perception of the earth inside is no idle poetry. Yet it seems to work with a lyrical pace, to be a good partner of an esthetics of place. It comes even before our ethical strictures about killing, and gives us those values. Like most of our deepest feelings, a sense of the earth inside is both survival and poetry, lyrical tough. It makes us care about life forms if we attend it.

16. Needn't-Running Rabbits

In those years which were my tenth and eleventh, I did not know how much earth I got inside me. It happened with a rhythm I took, runs and wanderings I did for no reason I knew.

No one told you such things as the goodness of ash for bows or worms for gardens. And, with luck, no one told you about a million years of tadpoles wriggling in your hand. Such things are perhaps untellable, except by how time tickles your palm, and you smile its secrets free, free as time itself when you are ten. At that age, there is a certain critical mixture of person, time, and place. For me, at least, it was young like that, at a bad time in the world, in a place with a small patch of wildness.

My family—then a mother, a sister, and I—lived in a house in Dallas just at the beginning of World War II. A little open land, in patches, lay thrown about the city, which was later to grow solid for thirty miles beyond its edges.

At our patch of open, a creek ran through near the house. Most of the creekbed was rock, the flat banks were rock. By the house, the bed was deep, with sides of a crumbly, flaky, light-rock sediment. Back of the house, the creek ran for a mile in woods upstream. On both sides, the woods went north and south for about a mile on each side. I knew one of those two square miles. It was a touchable fantasy, one I dreamed on and ran on. The white of that shaly rock became a quick, frozen moment I would remember forever. It was apparently of no consequence, just as when I climbed a fence at five and knew at the top while jumping that the second there would always be with me. There is no accounting for that. Solitairies, comfortable with themselves, are born in such moments.

From the house there on the creek, I could descend the dusty rock chips to the stream bed. I do not remember the name of the creek, though once years later I learned it. When I was there then, naming was too far away from me and from what I saw. We were raw and wordless to each other. Later, when I knew the name of the creek, I was not there anymore.

Going down to the creek, the flakes broke at my toesteps, and I slipped down to the sturdy bank. Hard rock. I had lived on wood floors and lawn and asphalt, gone up and fallen out of a child's number of trees. At the bank, I touched the warm and cool, the wet and dry rock of that stream. The bed of the creek slid down, algic and rilled with crystal water. There was a short, sandy-bottomed pool, deep to me, and then the flow was back on its shaggy green bed.

Late spring had worked the great surprise. Certainly it was millions. Frogs. Thumbnail size, they worked their vagrant ways in short hops at all the compass points. I did not move. Instead, I did everything else at once: I watched the frogs hop, I smelled the dank bottom of stream life, I heard a moving water somewhere beside me. I stopped still to take it in, and squatted slowly on the bank. To see the frogs, the newly bloomed tadpoles. I drew no conclusions from that moment. None. Yet my loyalties began there. To myself as a human animal too young to know any better. To the rock and water and frogs, for getting inside me.

I knew none of that. Hunkered down flatfooted, I picked a frog. Its wriggling in my palm made me set it free. In the stream there were tailed-and-legged tadpoles, still blooming in the sandy pool, wandering. Soon they would be out on the bank.

Around from the kitchen door of the house and out back, a path went off into the woods. It did not occur to me to wonder how that path had been made, who had passed back and forth before me, but I took it, through the woods. They were marvelous walking woods. Not much undergrowth, soft to the foot, with clearings here and there. The first opening on the path was a circle around an oak tree of ancestral size, casting cool shade.

There were always rings of charred stones around the oak. Inside them would be dank, damp ashes, flat in the circle; some light, dry ones would drift in the air as I poked them. Or there would be a smouldering branch, smoking with the smell of pork fat or a hobo's urine. And empty cans, still moist; pork and beans, tomatoes, potatoes. A few rags, a shoe, a flat, gone-out butt of wheat-straw paper on a rock.

I never saw anyone there, but when I got the way of moving quietly in the woods, I would look before I got to the clearing. No one. I looked around the tree. No one. So I made them up. Hoboes were exotic, about as outlandish as I could think of then. And the prospect of seeing "one" was fearful. What would "they" do? Kids are warned against hoboes, or were when there were hoboes. Later, when I was on the road, I knew better; but then, I made them up. I stared at those charred stones and saw men fat or thin, desperate or laughing grandly as saints ought to do, committing outrageous acts against the community, or only sleeping. Then leaving just hours before I got there, their spoor and legacy in a bean can.

That clearing was the largest I found in those woods, and I would usually take the path out back to there, and then go off looking by whim. Remembering, I cannot be sure of the connection, but it seems to me that if a day at school had been lively or if I got my home duties done quickly, I would run. Not just run, but dart and dash among the trees, into the brush, to a tree. Then stop. In a new place, by any tree, in the grass of a small, clear spot. And begin to listen, to watch and smell.

Perhaps the rabbits taught me that. In the times when I did not erupt on erratic paths, I simply strolled through the woods, on the worn ways. Or off them, slowly, bending branches aside, peering before, as I did not do while running. Those might have been times when more serious things were afoot at home. Or when the quiet took its way on me and I followed.

That was when I saw the rabbits. There were lots of them, cottontails, those startled, needn't-running rabbits. I would go along, quiet but not silent, and they would start up. Snugged down before I got too close, they would opt at some moment to run. Run full. When they went like that, I got still as a stone. I did not think about it, but I know now that I had to try to figure them out, discern them. Not wanting them to run, I did not want to catch them, either. They did not hop. Their legs pumped in blurred spasms, pushed off every tiny promontory they could purchase, and they whirred in silence off to safety. It was oddly sad to me that they ran. I wanted to hover over them silently to see what they did when I was not there, where they went when I was there. I did not hunt them or trap them. Once in a while, when I was still for a time, I would see one, short hopping and sniffing, nibbling, alert to what came on the breezes, and it would small-hop out of my empirical powers.

But when they started up and ran, I did not follow. Their eyes told me not to. As anyone—especially a child—would know, I sensed then all the lore we have about scared rabbits. In the running rabbit's eye is everything we know about fright. Every time, with each rabbit, that eye was on me. I only thought it need not run.

At times, the day and mood would take me on a path from the oak tree clearing toward the creek. Reaching it that way, the bed was higher and there was no steep side on the north. The woods grew, some sloped, and there was the bank. But on the other side of the creek, the south, the side went up high, broken by a gully. The white, fragmented rock walls came down in a narrow V. Inside the walls, erosion had carved a cave, but it had no top. Clearly, a cave was called for. A few times, I slopped across the water of the creek and sat in there, the vision forming slowly. Suddenly, one day I made a level floor. I was not neatening up, or making it comfortable. It was already fine enough. I was moving in, the way humans will do. Many of the trees around were cedars, and I used their boughs for the next step. The cave needed a top. Laid on thick, the boughs were a good roof. No one knew the cave was there. No one went to it.

It was "my" cave, cool in the hot summer. I went there more and more. It was about to become home. But I had one persistent problem. Scorpions. Almost every time I went to the cave, I found at least one light-brown scorpion. They were not much over an inch, but their reputation was enormous. As far as kids were concerned, they were deadly. So I lolled in my cave uneasily. It seemed large, but the scorpions seemed larger. Eventually, I only went now and then. But the path there was still familiar, and I kept it in my lore.

Often I sat on the north bank and looked over to my cave. I sat there and watched a lot. Mockingbirds. Trees on the banks. The sound and the flow of the creek, the bright chalk. In too much sun, I looked at the shaggy green of the creek. My favorite uncle had slid down from that part of the creek so often one day that he wore out his trunks, and his wife, my least favorite aunt, had thought his ass was offensive. No one else thought so, or even noticed, until she did. What I noticed was that he was a good man and she was a fool. Creeks can tell you things like that.

When I was alone, though, I lay in the shade of the cedars and looked across the bank. I got up and entered the sun, glancing down the stream, then up, then near, opposite, to the cave. You stare at your home, even if it is lost. And I felt no sorrow. After all, it was on the south side of the creek.

That had always been the forbidden side, where my cave was, across the creek. All kids knew that. I knew it, and it wasn't just the scorpions. That south side of the bank went up beyond anything you could see. Past that, no one knew what was there; no one ten or younger. I looked a lot at that bank. I think I tried to penetrate its mystery, to solve it, by looking at it, by making my cave in it. At times, too, we would climb parts of the south bank, and then come down from it. Once, from Friday to Sunday, two friends and I explored the creek upstream, walking bank and water, seining crawdads. We camped beside the creek. On the north side.

Yet no one ever talked about the south side. It was unknown, but we nibbled at it. The land was steep and rough and we did not know where it ended, on to the south. I think it only frightened us the right amount. Enough to breach it, all of us together.

Cedar Haven, the street in front of our house, came down a hill straight to the creek, then curved sharply to the left. Looking straight, before the turn, you could just make out the end of a road on the other side of the creek. No houses, just a road's end. It probably put the bridge in our heads and made the south bank vulnerable.

Our first try was a few downed trees that would reach across, and we tied them together. That bridge was close to the stream and a little funny because on the south side it ended at the bottom of a steep bank of black earth. It was easy to slide down once you got up, which just took one of those spurts of easy energy. In the winter it was a wonderful mudslide.

It was a surprise to us and should not have been when spring came and the water got the bridge in its maw. Our logs washed on downstream to a concrete bridge. Our downed trees were the principal players in a gorgeous log jam, which we regularly inspected for detained valuables.

We spoke of a grander bridge. Higher, obviously. That meant longer logs, tied tighter. And we decided on a railing. Once that was in our heads, we thought we ought to have a bridge opening, inviting our parents and friends from far away. That made the bridge different. We found tall trees, made them into long logs, carefully shorn and fitted. We talked about our bridge a lot to everyone around. We were the only ones in the neighborhood not laughing. But I should not let this bridge go too lightly. It was a structure, and it took time and care. We built it.

The opening was a Sunday. Our span was done. Far above the stream and its flood waters, the railed logs had a majesty we had not imagined. You had to walk it first. It had, to be honest, a little more bounce than we had reckoned on, but it was a sturdy waver, and there were the railings. All our parents came. I am not sure what we expected, but not what happened. They looked and they walked. And no one laughed. No one said how quaint it was. No one patted any of us on the head. After everyone left, I sat on the bank under the bridge and looked up, uncertain of what had happened. All of us, kids and adults, knew we had done something that was not idle, and we had done it on our own.

There was earth inside me. I had run with rabbits and seen their scared eyes. In those days, there were robins and bluejays, ash limbs for bows, and the hot, white chalk of the open places by the creek. They were part of me; I knew that. What I think means more to me now is that I was part of them. And I was alone a lot and had got those things inside. There is no other way to do it, but without the tadpoles and rabbits then, later the great blue heron feather, the feel of moving from the sun, the homing of the waterfall would have passed around me with no claims. We do not know how remarkable small events are. We do not attend them. We do not educate our senses or our feelings. And they are, anyway, always partly private.

What I saw, what I got inside, was wordless. I did not know what to call—how to name—all the life I saw and said breathless things about. No one told me. The rabbit's eye was as dumb as I seeing it, as starkly naked as the scorpion. My shadow, that of the earth inside, went before me and I did not know what hints there were in the small frogs I tried to raise but buried in the yard; in the cave; in the bridge; in the eye of the rabbit. I cast the shadow of a human animal before me, without a future I even thought of.

Not many years ago, I went to the place where those things had been my world. There was a bridge, higher and larger, of concrete and metal, with a roadway on it. It was directly over our bridge, which had long ago, I suppose, gone in tiny pieces out to the Trinity River and down in the Gulf of Mexico. I stood on that bridge and looked down at the stream. And I did not like it. Perhaps it was a kind of jealousy, I thought. A better bridge in a grown world. Or that a road and houses had pierced the south bank. Maybe just that growth by bridges and houses and roads does not cast the shadow I like.

I went down on the bank. It was tidy, but the stream jumbled by. I lay there and looked up at the road and knew why I was uneasy. I simply did not want to trade the rabbits and frogs and scorpions for a bridge. It was not enough anymore.

17. In The Superstitions

No one who has ever been there forgets. And I have discovered that to all veterans of the Superstition Mountains the experience is in some way unspeakably private and unique to each. There is a league of silence which these wanderers all join and all understand. It is obviously not the memory of silent mountains alone, but a tacit realization that not quite all can be said. Of course it may be rightly argued that these are people who by nature utter few words. They have fewer for the Superstitions. That is not to say the experience is opaque or even mysterious, but only that it is not clear why this silence is so.

I have speculated. Perhaps it is a kind of respect, a kind which grows from being lost, hurt, or terrified there. No one escapes them all. The reputation of these mountains as killers, maimers, and frighteners is well established. A riding group in Phoenix which takes parties on close calls in these mountains once hired a friend of mine to fire over their camp. They did not want the guests to miss the old prospector's rifle ousting intruders.

Or it may be that this shared silence of survivors is similar to the surprise of those who escape from undertows. You cannot believe the ocean can do that to you until it has done it. In the Superstitions there are other surprises. In almost fifteen years of hiking many kinds of lands, I had often wondered about people getting lost. It seemed odd. No matter where you are, if you can keep from doting on panic, your senses can take you out. About the Superstitions, though, people kept saying, "Yes, but it's different up there." So I went to see.

Like most people, I entered the country by way of Peralta Canyon. The way was fine and gentle at first, up and up, not very hot. When that kind of ease is on you, your senses wander from the few

feet ahead of your own feet. You look around more. The sides of
Peralta Canyon begin with great lumpy boulders, strewn beside you
by the chances of time. Up above, the monoliths sit, they slough and
ease slowly into this dust at your feet according to the gravity of wind
and rain. I counted great stony nuts, faces, tilted walls, fists, pillars like
guards, and then chuckled at my own human projections. Then I
stared hard to get at the time the walls had been there, and couldn't,
and chuckled more. There is no sign for that kind of time.

Certainly no one could get lost here, not with such characters
for guides. By evening I had changed my mind. Going down into other
canyons out of sight of Weaver's Needle, a volcanic neck and the pole
star of the Superstitions, the eyes' compass for these canyons, I began
to understand. They are all Peralta Canyon. Lost for a while, I laughed
at myself for doubting. Those same rock characters guide you everywhere. You must make careful choices. It would be better, if the
Needle is not in view, to be on the flat of the desert, where angles,
plants, far sights fix you in distance and direction. I had escaped being
lost for long, but I learned about recognitions which led to memories
and then to nothing but an utter destitution of place.

No, being lost is not the Superstitions' mark on me, but rather
something far more foolish. Gold. Everyone gets it there, the thought
of gold, gets it especially for the Lost Dutchman. No one knows where
Jacob Walz had his mine, but he said on his deathbed, the tale goes,
that you could see Weaver's Needle from his claim. I have always
suspected he was highgrading from some other mine, and deathbed
confessions can be last tricks. Being what we thought were sophisticated desert rats, two friends and I decided to forget the Dutchman
and look at the Peralta mine evidence.

Someone—the story goes that it was the Peralta brothers fleeing from Indians who didn't want them mining there—had inscribed
the full figure of a miner on a rock. Next to the miner is what looks
like a map. We reasoned that anyone in a hurry would not pause to
draw a full figure. So, we thought, perhaps the figure is the map and
the map a ruse. With a schematic drawing of the figure, we tried to fit
it to the configurations of a topographic map of the area. And found
one that fit perfectly. We were, of course, rich from that moment on.

Two of us had had experience in a real live mine, and we managed
to get the dynamite, caps, fuses, drill steel—all we needed to blast the

bejesus out of the hill we had found. We set out during Easter vacation to cache our supplies for a full assault in the summer after school was over. Now this place at Easter time can be hell on ice, with a hundred or more degrees during the day and in the forties or less at night. That daytime heat pushed us down hard into the trail. We shared toting a fifty-pound box of dynamite, three heavy pieces of drill steel, and six gallons of water, plus food and a rifle and other gear. My friend Dal was in his mid-forties, and Doc, the third, had asthma. A slow trip up for three fools.

When we got to the mystical site, we went off the trail as far as we could. We hid the dynamite under a giant slab of rock and covered it with talus; above it in a crevice of an even larger rock we put the drill steel and covered it. By this time the party was done in. That night we slept wearily, fitfully; and Doc was in bad shape. We were short of water so I left early as water fetcher. When I got down, the car would not start, so I hiked lightbacked to the road and down it to a strange little refreshment stand. They had no water. Only orange soda pop; bubbles in a parched mouth.

We did not return again until the end of summer, just Dal and I in 120 degrees of skin-searing sun. Because dynamite can separate in that kind of heat and be deadly to the touch, we went to set it off, remove the danger. Incredibly, the drill steel was gone. Had someone watched us? The dynamite was still there, and we dropped back about twenty yards. I lay down in the crease of a boulder and Dal peered around it at its base. With a .45 service revolver, I fired. Nothing happened. I fired again.

Today I only remember the flash, nothing about the *whump* of sound there must have been. The slab over the dynamite disappeared. As it was doing that, I must have turned and jumped behind the boulder. I remember crouching there as the slab rained on us. Dal was stunned and we looked at each other with unadorned disbelief. I was bleeding from face, arms, and back from rock shards. Dal was dazed.

I had violated what I had learned earlier, to go with the land and take it in, know it on its terms. And in the Superstitions. They are something like hot sauce. If you don't go with it and enjoy it, it can really hurt. I was up there for the wrong reasons, and the scars are there today. Of course, the Superstitions did not put them there. The silence comes over me, though, when I try to find out why I did.

18. Thales At The Flood

A lot of things happen at a flood. No one can ever keep track of them. Houses and cars and cats and tax records go all apart and float around and die and just get left where they are. No one wants to touch any of it. To have had three feet of water in your house is like having had a thief there. Your being has been violated. Smirched. Smeared. The water and the way it moves simply takes everything in the way of its path and of its speed. At will. Then it goes away. It leaves dead things and broken things everywhere. Most of all, it leaves silt in slits so infinite in smallness that no one can see it except by dusting every day. For years. It is not otherwise removable. Perhaps it never goes away.

When you first see what it has done to you, the hardest thing is to realize that just water was there. The depth of disbelief in foot after foot of water in your house, ruining every hidden crevice just by its touch, carries you, aghast, into residues you cannot take into your life. There is no readiness you can ever acquire. As with theft and fire, there is no way to accommodate flood.

There are certain chance distances one may have. My wife Molly was here in Pescadero for the northern California flood of January 4th in 1982. On January 2nd, I had gone into the hospital; some strange infection. On the morning of the 5th, I was still in intensive care when the nurse told me my wife had called and there had been a flood in Pescadero—and everything was fine. The news and its message from home hardly even registered. The next day it did. There was something serious at home. The extent of it came slowly, when Molly made a visit. She was tired down in the bone; she was distracted, still cold. She was huddled against some distant ravage. The hospital was not quite real to her: too warm, too dry.

She was still in the flood world. Just a few days before, she had spent hours getting things out of the house, out of the way, off the first floor of the barn, which is her gallery and studio, and my office. She could not do it all, even with help. Everyone was being evacuated. Clearly, she must go. With dog and one cat in the car, she spent the night in the high school with the rest of the townspeople. There was not a phone or a way in or out of the town, no electricity. And no food or bedding at the high school until much later. By the time disaster officials learned the town was flooded, there was no way to reach anyone. The residents were there over twelve hours, a short time compared to some people in other areas. But Pescadero is a small town, some 500 of us. Mostly we attend to artichokes, strawflowers, lettuce, and other coastal crops. We have redwoods and beaches. We have Duarte's, a fine restaurant with its own garden and fishing boat; a classic bank; a gas station; post office; general store; Norm's Market with fresh meats; a gift-and-clothes store with curiously good taste; and two art galleries with no tourist, kitsch, or hobby art. All on the flood plain of Pescadero Creek.

But coming home was no relief. The water had invaded, had been three feet deep, had left a foot of mud. Barely able to get inside, Molly opened the door finally and cried to see her place. It was a jumble of mud. Kitchen cabinets oozed the silty stuff from their farthest corners, in and over plates and bowls. The mud had to go, along with carpets and rugs, because of the health hazard. The firemen from the California Department of Forestry helped; the local fire department volunteers were up for hours, days. Neighbors helped each other and grew closer. Great as the temptation might have been, one did not just take a look and walk away.

Water, along with the sun, is one of the great shapers and movers of our earth and life on it. It is not only that we humans need it, other lives need it, and it shapes our land. More than that, it is simply pervasive. It is here. That alone has had plentiful effects all over and around us.

Thales, the Greek who is thought to be the father of Western philosophy, if there is such a sire, came down to us in history for one silly reason. He is the first recorded of "Western man" to posit one thing as the reality behind all appearance. His only dictum: Every-

thing is made of water. In the Idealist tradition, the most common
tradition of Western philosophy, there is some "reality" behind
appearance. In this case, "idealist" has no ethical connotation. It
merely means that an idea is more real than a thing, or even that
ideas are the only reality. To Plato, for example, tableness is real, but a
table is only an appearance.

Back to Thales, we find that there is one story about him, in
addition to his idea that everything is made of water. It is said that
Thales, while walking late at night and gazing at the stars, fell into a
well. He was saved. It might be up to an existentialist to suggest that if
one found oneself suddenly in a well, water might seem the only
reality. Which is only to say that we respond more to immediate
causes than we do mediate ones.

Water lives with us, and we with it, in both the close and far
ways. We need it. Lots of things do. We think water mostly just hangs
around for us to use. The distant shapes of water, though, make our
lives, drop to rill to creek to river to sea. Water carves our world and
pounds it away. It takes our lives and changes our lives. Most often
slowly, but forever.

No one died in Pescadero, but people still ask if life is back to
normal yet. I hesitate to answer. Life will never be back to normal.
Normal has been rewritten by water. Pescadero Creek is new in its
various banks. Roads, houses, floors were in over the gunwales,
sogged but unsunk. They are mostly fixable, and what can be fixed
cannot be mourned. It is a slogging, day-by-day, month-by-month act
to live newly, to adjust to, adapt for, to ignore from time to time in
order to live. "Normally." One gets too full of mud and dust and
mildew and soggy clothes and claylike gardens. There is nothing there
to suggest renewal. "Normal" has had no time to re-introduce itself to
anyone. We have to attend to the fact that such suspensions also
create a period of what seems to be a kind of exhilaration. It is known
to battle veterans and mountain climbers, among others, and it is not
the best period for keen observation.

Perhaps having missed the flood, I can take a more distant look.
Molly was always very wise. She did not try to tell me what our place
looked like.

The porch tilted back toward the house, which had settled into the earth south and east. The garden had a foot or two of new, oozy topsoil, the books were supine if they have been in the shelves under three feet from the floor. *The Meaning of Anxiety* had puffed to fluted edges, announcing itself with a flair for the moment. The dust, caked everywhere, was like flour or talcum powder, but brown. The mud-blackened bed and half the contents of the house were on the sidewalk, left for the overtime heroics of our trash collectors.

This had no connection to me at first. I was still on hospital time, which is to say, getting well. But I quickly forgot about that, in the way that important moments pass, and started to work doing things and getting dizzy and weak and worthless.

Molly had done or was doing everything she could. She had already gone beyond what I would have thought any human could do. I looked at her with curiosity and discovery. No one could have done all that. It was cold. It was wet. It was stiff. It was humid in the walls. There were worse places—Santa Cruz, Marin County—but I had not fallen in the well there. Or here. What if I had?

Nothing really tells you what to do. That is the key to all floods, perhaps to all disasters which touch human beings. Specifically, in your house, in your place, the choices are yours. They are by chance wise or foolish, taken at the moment, opted when the water is going from your ankles to your hips, and you take it as the moment to leave. The water might stay there; it might go higher; it might go down. At such moments, among others, you cannot know anything beyond your own fingertips. Some choices like that are clear cut, others are close, into seconds, and made or not by chances or wisdom or reflex. One by one.

What can be done about that? Being terrified, flooded, washed out, returning home or not, facing absolute ruin in the form of mud and water in every corner you knew or did not, looking every day at more and more mud and then dust and dust and more dust, finding week after week nothing but loss after loss, shoveling, hosing, cleaning, demudding again, having no electricity, no water, no phone, and over and over the same trek through all that, day after day, worming down to the tiniest needle full of Pescadero Creek mud, what would you do? That is not a theoretical question. In the car in the box in the bag in the case in the Bull Durham sack pinned on felt, I found the needle with its eye full of silt.

Any recourse is at least a way to fight back. Doing something about it is out there in the big world, not in the eye of a silt-rusted needle or the rotted contents of the house.

Here in Pescadero, the thing to do about floods started over a century ago, when the first settlers built houses on the flood plain of Pescadero Creek, as everyone did then, and many do now, everywhere.

Within our memories, the last great plan to fix the problem, with a number of hundred-year floods in the fifties and sixties and now the eighties, was proposed by the U.S. Army Corps of Engineers. The Pescadero Water Committee had gotten the County of San Mateo interested, and the county got the Corps' attention. The Corps of Engineers does nothing on a small scale. It proposed a dam on Pescadero Creek, which was referred to as "the Pescadero," as if it were the Mississippi.

Taking the Corps' figure for the dam's cost, there would have had to be 160,000 here just to pay for the dam. There were then, and still are, fewer than 7,500 people in this coastal area. The place has maintained an unusual balance of agriculture, recreation, and residential space for over 200 years, without one overtaking the others. It is a place which survives to this day as a prime example of that balance.

According to the Corps' report, there would have been no agriculture in the area by 1980. Naturally, if it took 160,000 people to pay for the dam, they would live in the most level spots first—where the farm land is. This in California, which loses 20,000 acres of prime agricultural land every year. For the Corps and the land developers, Thales was right. Water would have become very real. The old Greek ought to be the patron philosopher of the Corps.

The county supervisors did not go for the dam or any of the lesser plans in the report. The supervisors voted in the fall of 1970 to provide a flood control and water system here. Nothing has been done. In other areas where the flood rains of January hit, there may be questionable solutions to such problems as slides. Here, there was a start long ago. But we have no big names, no big money, no big vote, so the supervisors will probably go on doing nothing, as they did even in the time of a real crisis.

Which is what I did. Without knowing about it, I lay there and missed the Pescadero flood. Much as I wanted to, I could not go home at first. There, the bed and all the rest of the furniture lay, all soaked and squishing and splitting before Molly's eyes. She lived in the barn, slept on a mattress; no light or water or heat.

And I could not have done a thing, except for being there, to help. What would I have done with layers of mud on the deck, a film of dry silt in the house, crusts of mud in and on the barn-gallery, dollops and smears of it throughout my desk, a coat of ooze on the floor. A wetness hung in the air, wafting up dank odors from the underneath of everyone's nightmare.

It was cold. Rain flicked us off and on. I am usually careful not to give human traits to the elements. Yet there was the feel of something evil lurking around, in the wet, in the mud. It was the almost skin-tight closeness of a doom which had come and gone and left a smell of omen behind.

Things that dried simply dampened all over again; clothes were always wet to the touch. Every single thing, every fork, wall plug, bottle of aspirin, small treasure, had to be cleaned one by one.

While we were going through all this, and while people in other places were in worse shape, it almost happened again. On February 15th, we were up all night getting everything off the floor and upstairs in the barn. The rains had come back. By morning we had stowed everything. We waited. The creek had risen eight feet. It was one foot below the bank. In the night, I checked with a flashlight. In the day, I merely had to look out the window. By noon it had stopped. We left most things in a dry stash.

Time seemed a bit strange to me. I had experienced the result of a flood, then prepared for one, had missed one. All in the wrong order.

I missed the flood and yet learned from it the sense of disbelief everyone spoke of. There is no point at which one can say a flood has come, has happened, until later. Molly, in the real flood, had been walking into the house to get more stuff just before the evacuation. The water was up to her ankles. By the time she reached the kitchen, she was hip deep. Still could not believe it. She looked at the water and thought, This can't be happening. But the survival part of her dropped what was in her hands and directed her out to get the animals and herself into the car (one cat was gone) and to the high school, just as the firemen were checking all the houses.

Earth and water had a new dimension for me. I used to say, when it rained and some thought me daft to walk in it, It's only water. The "only" had dwindled in my mind, while the water and its silty cargo has grown. So, though I know the beginning and the end of floods but not the middle, I shall remember to believe the waters when they swirl. Or shall I?

One thing I know with certainty, and shall cherish as an historical and even historic correction. Thales was wrong. Everything is made of mud.

19. Green Hills and Dust

The Korean landscape as I knew it was both beautiful and horrible. The land was marvelous hills, green with shrubs, and the flat heat of the sun turned its soil to dust. There was a dust you had in your smell always. No one told you anything about Korea, the people, the country. It was as if we were there, sealed, and on the verge of being shipped from one limbo to another. But one lived in between.

The hills rose up from around Camp Casey, and we were not to go there. We were to live in the tents, four or more of us under each frame. After intelligence school in Japan, I was sent to G-2, division intelligence, of the 25th Division. We were then interpreting stereo aerial photos of North Korea in order to find out about troop movements and tanks and so on. The U.S. was not supposed to be taking such photos, and we were, as they say, "sworn to secrecy." Of course, the photo flights had been in the newspapers, thus creating secrecy and publicity, a matched pair for generals. The colonel there at G-2 did not seem to think anything was very serious, but he knew we had to do something, so we eyed the photos and noted the things we were supposed to, like people walking on dusty roads. I thought it was all very frivolous. It was.

One other thing we had to do was walk guard duty around the supply dump. The story was that North Korean and Chinese soldiers and South Korean civilians were raiding the dump. There had indeed been raids, and I wondered why a whole division of infantry could not come up with guards better trained in fighting off such raids. I got the feeling that the real troops were off somewhere doing more important things, like running in place. I did not know what an important event would take place for me.

Here it was. I had to walk one side of the dump, at night, from one spotlight on a corner to the next lighted corner. I had an M-1 and a clip with nine rounds of ammunition. That was all. I walked back and forth. One begins to think. What if anyone tried to break in? I probably would not know it, because I was a perfect target there in the light with night all around. But what if someone did come down for the fence and in?

There are certain moments when there is no more room for any decisions, just as with a flood, when you act or not. I had to know, from myself, what I would do. I remembered Doug Wesson telling me that handguns are always and ever used for one reason only: To kill people. And that was why I had nine rounds, which is a paltry sum of killers for a guard. I was obviously not important in the grand scheme of things, but I thought that if anyone heard rounds going off, there might be some movement toward where the sound came from. It strikes me now as amazing that wars and other human excursions are announced, discovered, and made fatal by sound and noise. As much by the human voice as by the human bomb. Even ancient, silent problems become crises just because some president or other makes them so; down in news or up in news.

I walked. One light to the next and back. It occurred to me that I had to decide what I would do. On my own terms. And I chose the idea, the fact, that I would not kill anyone. I stopped and sniffed the dusty air, and knew I was supposed to kill anyone who came for the fence of the dump. And I would not.

At this very moment, I do not consider myself a pacifist, whatever that means. I know I would kill in self-defense or if anyone threatened the life of someone I loved. Once, I came within inches and moments of doing just that to a psychotic who threatened my wife. It was a near thing, with the kind of horror of fear added to that of knowing by your own example you are capable of killing. It was just that walking that guard duty, I had decided I would not kill anyone. Like everyone else, I knew the primeval urge to kill, but never had. It is more difficult to get familiar with the feeling of not killing. That urge is just as ancient for the human animal as the kill. Oddly enough, though, the capacity not to kill is not as acceptable in this society as the willingness to kill. One must only admit to the popularity of war.

No one can tell you what to do in a situation like that night of guarding the dump. You make what is an emotional and philosophical decision, at one instant, and your life goes that way. You live your future as if you meant it to be that way. And you did, in a kind of unattended moment. I find that odd because I cannot believe that people plan their lives. And indeed they do not; they assume them. School. Job. Marriage. Kids. Retire. Die in conclusion. If you are cursed with the idea of choice, the whole layout is off center. But what happens if you are already in it and find that you choose against the pattern?

You don't shoot. Lots of people in the army don't shoot. The reasons for that are endless. You can't even make a patchwork quilt out of them. Perhaps a Navajo rug, with a place for the soul to leave. This is how it happened for me.

Before I walked the guard duty, I walked the hills. One Sunday I decided to spend the day hiking up there in the green hills and dust. There was really no reason not to, but no one did, so I knew there would be some solitary hours as I hiked. There were those hours of solitude, and there were not.

Alone, yes. But flooded with the human presence. Every single square foot had some reminder of battle. Shell casings. Ammunition boxes. Discarded clothes. A fragmented boot. Whistle. Belt. Sock. Weapon parts. Foxholes. Helmet. Snapshot. Endless shell casings.

At first it was a matter of wonder. The artifacts of battle. The last story of so many people. For hours I walked on. The same, unremitting shards of war. And suddenly the waste overwhelmed me. Human animals had concentrated on killing each other, had killed many other animals, had ruined this and many other habitats. For the first time, I saw that war is—and I did not say this in my mind—an environmental disaster. Everything goes. People, other animals, plants. For the greater glory of—nothing.

We tend to separate ourselves from the other animals and the plants and the very land we walk on, and war is no different. We think of human casualties and the destruction of things we have built. We even make the distinction that it is heroic if young men are killed but not old ones or women and children. The rest of the waste is just as serious, but the truth is that we do not care about it. Just as we do not really have any sense of future debt, though we mouth many things about generations to come. The way we act, though, is like this: Get yours, now. The rest is silence.

Indeed it is. Nothing moved or sang or grew in the DMZ. Now, it is said that many species are coming back, some new ones getting established. This is supposed to be a good side of war, keeping people away from the land. That may be a good idea, if we could do it without war. A depeopled zone. A museum of the earth.

That walk may have done it for me, made me decide not to kill. There on the ground, littered at every step I took, were signs of the death we had done to each other, to our earth. And it cannot be made up for, there is no retrieving it. A hill here and there, glorious deaths of a magnitude we do not even imagine.

On that walk, I felt the need to cry. I think that is a bad sign. One ought to cry or not. To feel the need is too civilized, and so much the young age I was. At the sight of that destruction, with its message to me, all I could do was sit and let a profound depression wash me around in the dust. I felt a kind of inarticulate ignorance, the intensity of unnamed discontent. Hour after hour, all day long, I had seen this heritage of waste. Because I was truly ignorant, had never seen such things, known or read about them, I was in shock, unready; in the silence.

Walking back to Camp Casey, I knew that I was in some way changed. It was more than my life could define, could say, could get over. I was unable to talk to myself. Or anyone else. It made a kind of distance I could only later make up for by talking to other men who had felt as I had, felt alone in that knowing moment. It was not the kinship of battle. It was an unspoken sharing in the kinship of loss.

It seems that what the gifting birds left in these cases is characterized more by paradigms of feelings than by some specified, well defined content. Perhaps impossible to describe, the feel of the earth inside can be expressed, conveyed with the oblique clarity of side-lit images. As Bergson found out walking around his bell tower and William James discovered walking around a squirrel walking around a tree, one encounters a new frustration in realizing the entire thing is never seen. The description of the earth inside is newly defined on every occasion, and never complete. Definitions and examples can be interchangeable if one accepts the possibility of uniqueness.

Like many highly valuable feelings, the sense of earth inside is a lasting, personal one. The discovery of it, such as in the rabbits' woods, is as exciting as is the sense of its loss or violation, which I experienced in the Superstitions by my own hand, and in Korea, where I walked among the artifacts of the dead. And certainly no one's examples of the instants of the earth inside would be the same as those of another. Thales at the flood could have said other things to me or to someone else. From it, for me at the time, I got the sense of neither surprise nor anger at the physical power or the power over the mood of man contained in that invasive combination of water and soil. That is one of the ways earth and people are, faced with the awe of the wriggling tadpoles and the irresistible waters. One was a discovery, one a disappointment, both of which are values we have created on an indifferent earth where everything is equal. Except to us.

Perhaps these experiences come as esthetic respites from another attitude toward the earth, that of using, conquering, consuming the earth for our own survival and well being. That is not to say that esthetic events have no use, for nothing better fuses the mind and the senses, the individual and the rest of his world. The ability to make that combination, and our pleasure in it, is one of the defining characteristics of man the animal, seen without shame or pride.

The sense of the earth inside is one of those experiences man can see at his fingertips, with the attention of what might be called a mind only slightly sprung. It may be that point-of-view is what is needed, also, in looking at the other animals and the plants as if they were members of the family.

VII

FAMILY PORTRAITS

We have that sense of the earth inside, though we hold it away, beyond touch. We hold our sense of community with other kinds of living things even farther away. It makes no sense (except that man is the categorizer here on earth) to separate history and natural history. The events of man's history and that of the other animals and the plants are literally inseparable. Our food, our protection from all the weathers, our arts and medicines and pleasures are mixes of the histories of man and the other lives on earth. There is no honest historiography without a picture of the paces we take with the egg or the potato, leaving out wool and cotton, excluding birdsong or the plague. And we ought to but do not take special care because we alone can create or destroy the others at will, and therefore, at some point we do not know about, ourselves.

Yet we will on one hand say in our formal moments of knowing better that man is one of the animals. But not that we are animals. Not that I am an animal. No, we mean to degrade by "just an animal" or "behaving like animals." In those instances, we take ourselves to be higher than or outside the class of animal or plant. But then, it might be said, man is the animal which ranks things. Naturally we know who will come out best, on top, or even in a separate class.

There are, of course, those who think man is worse than at least some other animals, that he is the defiler of the earth, the spoiler. Just because man tends to be a meddler, not knowing when to stop fiddling with things, his lack of knowledge becomes either quaint or deadly. One finds this attitude among some conservationists and naturalists, the latter of which write about all kinds of animals. Except man.

It is a simple truth that the other animals and the plants are our companions living on earth. But like the sense of earth inside, it is one thing to know that and another to feel it. To make that emotional recognition, it is not necessary for otters, say, to become cuddly, just as viruses, from this point of view, need not be evil. After all, one need not feel the soil between one's toes in order to sense the earth inside.

Such things are man's hyperbole. Man has certain characteristics which set him apart from all other creatures. The same is just as true of every single other species. And every individual. Still, we have shared histories on the earth.

20. Home With Daphnia

The discovery of daphnia came for me early one Sunday morning while my wife and our house guests slept. I watched daphnia for hours. Peering into her world, I did not at first feel I was intruding. She and her friends swam at random, free as daphnias are. In their liquid home, giant green and dark brown specks floated, migrating slowly around like icebergs in the salt sea. The motes of that world I saw were flicked in the backstream of daphnias on errands I could not even guess at, but I placed a bet on food, with its debris in their wake as they roved. I could not be sure yet. It was my first look at their place.

It is always a surprise, that first look at a habitat under the microscope. The slide tells you little, usually nothing, as you hold it. You hardly look, barely attend. Get the focus, though, under the scope, and all is changed. Live beings on an unprepared slide move by your surprised eye in colors as florid as purples and yellows. The red and blue dyes used to stain an organism, or part of one, for looking, create curious forms of art which momentarily distract a mind not dedicated totally to science. A microtomed house fly, section by section, gives up many new secrets. The fact another world lives down there is not trite if you take it quite literally.

In this case, with daphnia, the pond water had looked clear to the unaided eye, only a little speckled. One drop went on the slide; on went the cover slip, a very thin, clear plastic coin used to top the drop that was her home. I peered. Under the scope, the furniture of daphnia's universe began to look overstuffed, as teeming as the living room of a recluse. And yet these tiny water fleas, crustaceans just a millimeter long, made their way about with only two guides. Their first antenna is a chemoreceptor, analyzing the recipe of the water around them, leading and turning them. The compound eye, each lens bulging alone from the cupped recess, vibrates rapidly, coordinating with the chemoreceptors in search of food and flight from danger. With the second antenna, daphnia pulls its way around home, locomotes smoothly in a world silent to us. When the food is found, five pairs of feet simply catch it, refer it up the body, push it in the mouth. From there, into the midgut it goes.

All of that is marvelous enough, but there is more. Daphnia is transparent. Extraordinary, to see right through it. Down there, in the microcosm of the scope, in the water littered like a beach after Labor Day, is that pinpoint of swimming light. The first look stunned me beyond curiosity. Nothing about daphnia mattered except that, in disbelief, I was looking down into it, not just at it.

The fifteen segments of the carapace, the brood pouch, the very head itself—all clear as glass. You can watch its heart pump the blood (at a pulse rate I later counted as 180) without arteries or veins, see the corpuscles course over the brain, down the belly of the body, back around to the heart. As the animal lives before your eyes, you can see the brain. The optic nerves unite into the optic ganglion, all visible.

Daphnia is a tiny relative of the crayfish, a miniature crawdad not very much like its cousin, but with a family look about it. This daphnia, one of many varieties around the world, lives a mobile life in a tiny niche of the plankton world in ponds and other open, fresh water. Fish love it. Another family member, *Daphnia magna*, is three times as large, and many are raised for fish food in pet shops. The smaller daphnia I focused on in the scope is eaten by fish only in its own home, and is otherwise mostly ignored, left to its own tricks where it can grow, where the water is fresh enough, where man leaves it alone. Biology students do meet it occasionally. They see it somewhere in the course sequence between the crosseyed *opera bouffe* of the worm planaria, and the gleeful frog, *Rana pipiens*, which has started, with shrieks, many a lab feud and romance as well.

Such students who see daphnia then expand their horizons to Government 101 and Philosophy 1A, forgetting daphnia altogether. It is just as well in our universe. Things useless to man survive, and it is only a matter of pleasure that daphnia has no dedicated followers, no commercial need for its body. Otherwise, it would hit the passenger pigeon trail. For me that would be an esthetic tragedy beyond coping with, as well as yet another small gap in the food chain, which would hurt us all. It would, in fact, be the end of one of our secular miracles.

As I looked in on daphnia, I suddenly found one with a full brood pouch. Most daphnias are females, and they reproduce by parthenogenesis, or cloning, without fertilization. If the environmental conditions are right—all females. If it is perhaps too cold or there is not enough food, some of the young mature as males. As I watched, my daphnia was struggling to push the young out. And they were trying; you could see them clearly, exact but smaller copies of their parent.

Like a lunatic, I shouted up from the basement, "Daphnia is giving birth!" All were awake upstairs by then, and came running down to see what deranged mission I was on. I explained quickly what was in the scope's field. Carefully, everyone looked, cheered the mother on, put their own maternity and paternity into that drop of water. They could *see* it all.

After they drifted back upstairs to await my announcement of the birth, I wondered if this fight were normal; it did not seem so. Then I noticed the mother was near the edge of the plastic cover slip, barely pushing. Could the cover slip be weighing on this one? With extreme care, I used a fine needle probe to press on the slip, slightly bearing down on the edge opposite the scene of birth. Immediately, three young popped out and swam away with no effort, free as daphnias.

Without qualification, I can say it was a feeling of great wonder to be a midwife to a daphnia. Though without my cover slip she may have needed no help, no one to attend her giving of birth, I had made up for interfering, I hoped. The young swam away in health, searching for food their feet would carry where it counted, for life outside the pouch. I kept my eye on the mother, sluggish in her moves, to see if she would recover. In less than a minute I could see she would not, that she was dead. Perhaps, I thought, that happens to all of them after giving birth. I later learned that it does not. I had seen an event. Made of my meddling or the last strength of one daphnia? I lowered the slide back in the tank of pond water, putting them all, the new young included, in a vastly expanded world, until I later placed the pond water back in its home, in turn. Then I went upstairs. I have not seen daphnia since.

21. The Surprise Of The Dawn

The bark is a smooth, rilled red, and thin. Perhaps because I knew the tree came from China, it had a delicate, Oriental look, as if mist should be its corona, the light of soft afternoons its aura. When I first looked at the dawn redwood, it was a moment in which I might have become a mystic. Certainly it would have happened if I, like Wang Tsang, had seen over a thousand of the trees at once and known there was nothing like them in the world.

The dawn redwood I saw, by itself and known for what it was, even had a label on it. An old, handwritten card with plastic cover read, "The Dawn Redwood *(Metasequoia)* from China, not quite a year old, August 20, 1949." But Wang Tsang had seen the dawn first in unadorned and undefined surprise. In 1944 in the Hupeh village of Shui-hsa-pa on a forest survey, Wang noticed the tree and could not guess what it was; he knew it did not grow anywhere else. He sent needles and cones to the National Central University in Nanking. No one knew the tree and the samples were sent to Fan Memorial Institute in Peking. Dr. H. H. Hu was startled into disbelief when, the cones and needles before his eyes, he discovered they were identical to fossil remains of *Metasequoia glyptostroboides*, the dawn redwood. It had only been recognized as a separate, fossil redwood species by Dr. Shigeru Miki in Japan in 1941. The dawn had been extinct for 20 million years.

News of that kind comes rarely in the life of a paleobotanist. It is like a paleozoologist hearing of the discovery of a living dinosaur. Indeed, when the dinosaur lived, it was the largest land animal and the dawn was the largest plant, but that was 150 million years ago. At the University of California in Berkeley, Dr. Ralph W. Chaney, leading expert on living and fossil redwoods, heard about the dawn and knew he must go to China to see the trees, known in Shui-hsa-pa as the water pine.

In 1948, Chaney and Dr. Milton Silverman, a science writer for the *San Francisco Chronicle*, flew, boated, rode in sedan chairs, and walked to Mo-tao-chi, in Szechuan Province, 70 miles from the Yangtze River, through rain, bandits, mud, and mountain cold. There were three trees in Mo-tao-chi. The largest was 98 feet high, almost 11 feet around at its base. From core samples, Chaney estimated it to be 500 years old. The people of Mo-tao-chi told the party there were more trees 40 miles away at Shui-hsa-pa. Chaney, along with Silverman, a translator, a forester, and several "bearers" faced the trail again.

Shui-hsa-pa was an experience Chaney and Silverman would never forget. The valley had only been settled for 200 years, with nomads passing through before that. There were 1,200 dawn redwoods, mixed with oaks, maples, and chestnuts, with azaleas, hydrangeas, rhododendrons, wild strawberries, and bamboo. Surely it was one of the oldest surviving ecosystems in the world. In the red soil there were reptile fossils beyond counting. Chaney said he would not have been surprised to see a dinosaur come foraging along the trail, so ancient was the feel of the trees and the red earth filled with remains of plants and animals. This was the only stand of dawn redwoods on earth, and it was a 20-million-year-old surprise.

Chaney told the villagers about the trees, but the head man said any effort to keep them safe would have to come from higher officials. His people would always need wood. So Chaney made a plea to those officials on his way home, but no one has been able to find out what happened to the trees.

In one of the least-known successes of the return of an endangered species, a very rare one, Chaney brought back four seedlings and many seeds from his trip. Plant inspectors in Hawaii could not find any regulations against importing the tree. Like most people, they had never heard of a dawn redwood. So the trees got through, and because of Chaney's efforts, it is now estimated that there are more than 125,000 dawn redwoods growing outside China. Most are in Japan and Korea; there are many in California, Oregon, and Washington. In much of the United States, as far east as New York, the dawn is sold as a "dwarf redwood." Growing at a foot a year as young trees, they are hardly dwarfs, though they are shorter at maturity than the Pacific coast redwood *(Sequoia semervirens)* and far less massive than California's giant sequoia *(Sequoiadendron giganteum)*. And the bark is thinner on the dawn than on either of its descendants. Unlike the coast redwood or the giant sequoia, the dawn loses its leaves in winter; it is a deciduous conifer. Many people think the tree is dead when it sheds its leaves. It is only going into hiding, waiting for the spring of Shui-hsa-pa.

The first dawn redwood I saw was one of the seedlings Chaney had brought back 35 years before I saw it. It grows in the yard of a house once owned by friends of Chaney. To another friend, Donly Gray, Sr., a botanist and nurseryman near Sacramento, California, Chaney gave many seeds. For the rest of his life, Gray planted the trees and gave away seeds on the West coast from Mexico to Canada. The dawn was on its way back, growing again where it did millions of years ago, as an ancestor of today's redwoods.

Almost 30 feet high, the tree I saw before me demanded that I stay with it for half a day. The red bark, the filament leaves, the heavy branches, all wrapped themselves around me like a lost friendship. For a year I had been trying to get a visa to go to China and see the dawn, if it is still there. Finding this one was a wild chance. I had told a friend of mine, visiting from Hawaii, that I wanted to do a book about the dawn in China. When he left my house he went to a restaurant in San Francisco, where he happened to meet a photographer and mentioned the dawn. The photographer knew the people who had planted Chaney's tree in 1949. It was eight miles from my house. That led me to Donly Gray's son and his nursery.

For the dawn, it had been a long way back from the edges of time. From the first short mention in a book, my first look at the tree had been chance threads tying me to it.

It is difficult to explain about the dawn. I am as fond of trees as I am of people and other animals, but not a tree zealot. This one got me. A delicate, strong survivor among us after what was thought to be a 20-million-year gap is a curious thing to see. That amount of time is, of course, an absurdity. I know of no one who can get 20 million years into the gut, much less into the head. It does not scan, that poetry.

I look at the tree, growing with no special care; growing where the seasons are opposite from its home, here with rains in winter, dry summers; growing with solid health in its gray-green cones; growing as impossibly as a cloud with a twin. I touch it. I stare at it. I walk around it. It is, after all, simply there, without any fuss of its own. But I can't help it, though I know, as with love at first sight, I understand that tree no more than it understands me. Perhaps we are both mystics.

22. The Otter Stone

For any sane sense of reality, the otter stone is the one used by that marine mammal as anvil and occasionally as hammer to open its shellfish meals. That is an otter reality which has not changed in the last twenty years, but the metaphoric dimensions of the stone's use have taken on a locus more similar to a battlefield. The Friends of The Sea Otter want to throw the otter stone at the friends of the abalone because shellfishermen don't like otters. And the abalone fishermen want to throw stones at the otter. And the otter just keeps on eating its way north and south from its pre-1960s enclave at Big Sur on the coast of California.

If one wants to speak of residence on the California coast, the shellfish have been there over 100 million years, the otter about three million, and man for something like 30,000 years. All are natural to the area, having been thrown there by the fine tuning of the universe. Aboriginal people were able to kill the otter for its warm pelt and for food, and they competed with each other for shellfish. Even where the otter was not hunted heavily, ancient man used wooden wedges and whalebone prys to dislodge abalone in crevices otters could not reach.

Neither the Indian nor the otter were to last for long. In the otter's case, Russian and British fur traders began to hunt it before 1750, but it was not until the efficient Yankee with his armaments that the animal began to feel the pressure of the hunt. Between 1741 and 1911, a million pelts had been taken, and the otter was thought to be finished. It seemed to be too late in 1911 when an international moratorium on hunting the otter was declared. Estimates today of the number of otters left in California at the time range from 30 to 100. They lived in almost secret survival at Point Sur, where the lighthouse keeper in 1915 counted 32 animals. A California law has

140

protected the otters since 1913, and the federal Marine Mammal
Protection Act of 1972 included the otter, taking it out of the control of
the California Fish and Game Department.

An indication of how little is known, or perhaps in this case
"decided" is a better word, about the otter is that it is not settled yet
whether there are two subspecies of sea otter or three. The animals of
the Kurile Islands are classified as *Enhydra lutris gracilis*, while those
in Alaska are *Enhydra lutris lutris*. For the California population, the
otter stone gets hurled around a little more. Some say the California
otters are a third subspecies, *Enhydra lutris nereis*, others that the
southern otter is the same as the Alaska one. There are similarities
and differences, both obvious and subtle, based on anatomy and
behavior, but no answer based on such things as definitive blood
studies of the 2,000 otters in California. This is more than a casual or
even objective stroll down the arcane halls of taxonomy. If there are
2,000 of a different species in California, they qualify as members of
an endangered species. If they are the same as the Alaska population,
there are 102,000 of that species. The isolated otters in California are
endangered or not depending on our definition of them. There is,
though, sympathy to include them as endangered because of the real
possibility of oil spills in their range.

The otter is, of course, oblivious to all this, moving along the
California coast at about two miles a year south and one mile a year
north (there are more of the otter's favorite foods in the north, so it
lingers longer). The exception to that movement was 1973, when the
otter path happened to cross sandy-bottomed ocean in both directions, moving in that year 18 miles south and 14 miles north.

Living in a range from Santa Cruz in the north to just south of
Pismo Beach in the south, the otter owes its survival and burgeoning
population in part to being left alone by man, though some otters
have been found shot to death. But the main factor in the blooming of
the otter is that for a marine mammal it has remarkable adaptability.
Females can give birth on land or in the sea, adults and young can haul
out of the water at will or not, they can live among kelp beds or not,
and they eat an amazing variety of food.

The otter menu includes abalone, mussel, sea urchin, crab, clam,
squid, limpet, chiton, barnacle, starfish, and scallop. In California, the
otter's appetite, timed by a metabolic rate five times that of man, leads

it to consume about 25 percent of its body weight per day, for an average adult of 55 pounds. That adult will eat two and a half tons of food a year, which it forages for at depths down to 145 feet, though it usually goes no deeper than 120 feet. If the present population continues its spread up and down the coast, the otter in California could reach 16,000 in number, eating 40,000 tons of shellfish a year. Man's commercial catch in that coastal area is 4,000 tons a year. Those who go to sea for the largest fisheries—abalone, crab, clam, sea urchin (the urchin is mostly exported to Japan)—have cause to worry about the otter taking their mode of making a living away, for while the otter does not eat any of the shellfish to extinction, it gobbles away the sport and commercial fisheries.

Though the otter gives birth to only one pup every two years, its food needs are quickly taking it along the coast. When an area is almost depleted of shellfish in otter numbers, a small group of mostly young males, called a migrant front, moves to the next place with nourishment. Ahead of the front are advanced foragers, fewer in number, and beyond them are wanderers, which have been seen as far as 400 miles beyond a front. The otter raft (otters come in rafts the way lions come in prides) is a group of determined settlers, moving on only when the food supply gets too low. The most common cause of death among them is starvation, and it is not clear yet what other dangers the otter faces. It is run down by boats and shot, which accounts for 15 percent of otter deaths. Teeth of the great white shark have been found in otter bodies, but man has destroyed the land-based predators which once attacked otters. Still, the otter does not haul out as much in its southern home as in its northern site, perhaps because of warmer water and the presence of large kelp canopies for rafting.

The lines drawn here limn a perfect definition of the classical environmental confrontation. A healthy, bursting population of an animal which is considered useful to man for its pelt is also a threat to the livelihood of some and a darling to others. And it is an animal about which man knows very little. Its fate is clearly in the hands of people, as is so often the case. What is to be done?

The shellfishermen and sport divers and foragers want the otter contained, at least. The abalone divers seem to approach the otter with the most hyperbole, apparently wanting to drive it off the edge of the continental shelf. Stories of shootings run their gleeful rounds among the divers. One said he has a dream about being out in a boat and shooting otters until they are all dead—and he then jumps into the ocean. Possibly a somewhat guilty sort of dream's end.

Mariculture has not yet succeeded for abalone, and there is no doubt at all about one thing: Where there are otters there is no abalone fishery. Entire familes see the "ab" about to disappear, along with the divers' boats, homes, way of life. The crab and clam and sea urchin fisheries would also cease to exist, and, understandably, shell-fishermen want to see the otter in another place.

The Friends of The Sea Otter, however, have a different and often more flamboyant point of view. Said a letter writer in *The Otter Raft*, a Friends' publication, "Abalone? Ah baloney!" To many, also, the otter has a very high cuddle quotient. Small enough to be a furry pet, it does such things as turning somersaults in the water, which seem playful. In that acrobatic move, however, the otter is blowing air into its fur, for it has no blubber for protection. The most obvious human connection is that the otter is a deft tool user. Catching a crab or clam or other delicacy, the otter surfaces, places its stone on its stomach, and brings the shellfish down on the stone, using its agile front paws. It is thought that the otter also uses the stone to break open shellfish, especially abalone, underwater, where it also uses its paws as prys.

To the Friends, the shellfishermen are predators, marine machos who even stab otters with knives. And the fishermen think the Friends and the Sierra Club as well are taking away a living just because they think an animal is cute or beautiful. So far, neither the state nor the federal government have come up with a "management plan" suitable to all sides. No serious thought has been given to the alternative of "harvesting," that is, shooting, otters as they get beyond certain limits, but investigations have been made into some kind of manmade birth control technique. There are no answers yet. It is known that what happens to the California sea otter is up to us.

It is unfortunate that we must always think in terms of what use or threat an organism is to man. We need not, for example, be concerned about redwoods just because they are either beautiful or profitable. They are worth our care simply because they are red-woods. In the case of the otter, we ought to be able to care for it because it is an otter, whether it is cute or threatening to people. We are so fond of thinking we are outside of nature that we do not stop to think that if that is the case, nature is outside of man, which is just as much a delusion, but might give us enough pause to understand that the otter only cares about otterness, and should be considered on those terms, as otter.

We are supposed to be more enlightened and unselfish than simple thoughts of use. Such a judgment is merely the hope of potential. The otter needs shellfish just to live; we do not. Ironically, it is the tool the otter uses which endears it to us while it is at table, and threatens us while it is gone to the hunt. It does not know the blessing and the possible curse it clutches in the otter stone.

23. A Question of Pelicans

In the small bay at Zihuatanejo, the pelicans fed and flew all day. They floated, seemed to play and preen. Some days I watched during all the hours of light. In their flying, *los pelícanos* look like two animals pieced together. The back, the wings, are grace and ease, but the head is tacked on, the beak outbalancing frontward so that any moment there will be a somersault. Few birds fly closer to the surface of the sea, and that great cocked head with its Cyrano beak preceding it looks like impending disaster. Fish found, the pelican climbs, then drops like a dart. Another sea bird, the gull, is smooth and of a piece in flight; but when the gull dives, it appears to fall, lumbering toward the waves out of control. Pelicans, when you first look at them, seem unfit to fly at all. Yet, diving, they are one bird. Before it hits the water, the pelican's wings go back to a V, the head streaks forward, all on target, aimed with its own sense of hunger.

Zihuatanejo is about 145 miles north of Acapulco on the Pacific coast of Mexico. Twenty years ago the last twenty miles of road were unpaved and no cruise ships stopped there; but yachts did and turtle factories afloat did, and fishermen came. The harbormaster there was as often in a plane as in his office. He soared up from the small airport above town, a .45 in his belt at all times. That was not unusual. At a large wedding in the hills, almost everyone carried a .45 and shot it to the sky like fireworks. Some of the people at the wedding had been in the hills when linemen were trying to bring electricity north to Zihuatanejo. They had shot the linemen off the poles, and when the *rurales* stormed the mountains, they shot still, especially aiming at those who had "shiny things on their shoulders." There were still no phones or power lines there, just generators.

Down on the beach there were five hotels, with two on the cliffs

above. Those two were the civilized places; one offered "safaris" to
the hills and the other had a cog railway to the beach and back. I picked
a beach hotel with three rooms and a thatched roof in front with six
hammocks under it. I planned to learn how to lie in a hammock for at
least one day, just watching. There would be breaks for the incredible
meals from the sea, and one more before dinner, when a boy wheeled
a barrow of coconuts by. I punched the one soft eye, drank, added
brandy, watched the pelicans.

An instant after a pelican hits the water, it pulls its beak up, flops
its body back head up, and shakes its wings a little. Then waggles its
head to down the catch, and is off again. Its feet waddle, pushing off,
synchronized with the wings. When gliding to a landing if not diving,
the feet help too. They put down to skim the water and slowly go
under. My one day turned into every other day of watching pelicans.

Next door to my hotel, a middle-aged Mexican with a young
Asian woman and an entourage appeared. His yacht was anchored in
the bay, a .50-caliber machine gun mounted on its stern. The man was
a millionaire and had his own army somewhere in Mexico. Called
"Los Dorados," they were going to take over Mexico. As he told his
story, it turned out that he was born north of the border in New
Mexico, had been orphaned and then raised by a Japanese family. In
World War II, he had captured 2,000 Japanese troops, including a
general or two, by simply going up to caves on Pacific Islands and
telling the soldiers they were surrounded. By himself. He had had a
medal taken away for shooting a Japanese officer. When a movie was
made of his exploits, he was told he could not play the part because—
he was not the type. Bitter, he took his money to Mexico and made his
fortune. In Zihuatanejo, he fished from his yacht and went out fishing
with Gregorio, the town's genius fisherman and boatmaker; more
than that, Gregorio was an unrecognized mystic of the sea, the kind of
man who sits out every dance but listens carefully to all the music.

One day when Gregorio and I were eating *huachinango*, the
great Mexican red snapper, a giant blond man came dripping from
the sea. Gregorio had seen a yacht drift in the night before. The man
was captain and crew of *Quan Yin* (the Chinese goddess of mercy), a
64-footer without a mast in the bay. The owner of the ship, sailing
literally out of a book, had started from Boston with a crew of five as
green as he. Barely surviving a storm, they returned to Boston and

picked up this captain, 20-year-old Rick from Nova Scotia. Rick was equal to the problems until a storm off Tehuantepec took the mast and sails off to sea. The ship was towed to Acapulco by an Argentine submarine, and everyone left for Boston, leaving Rick to take the ship to Los Angeles on a skimpy diesel at five knots. The engine broke down for the last time just before Rick drifted into Zihuatanejo. By telegraph, he notified the owner, who had parts sent from Los Angeles. They were the wrong size. Rick went to Mexico City and made the parts himself in a machine shop.

Before Rick left, his diesel running with a purr to Los Angeles, the commander of Los Dorados had a grand fiesta for everyone who wanted to come. It took place on the shore of the bay just opposite the hotels. You could take the long walk around or Gregorio and other fishermen could ferry you across. The harbormaster was there, Gregorio when the trips were done, Rick, the beachcomber Bo, Ed and Harry from one hotel, the couple from Los Angeles from another, Dante Piccolo from Chicago. As if paradise were not just made of palms, sun, and warm waters, it seemed to need all of these people, about 50 in all. There on the shore, with just a poled roof nearby, we saw the day's festival before us. There were snorkel face masks and Hawaiian slings for skin diving, cold beer, spiny lobsters, and— cooking on racks—pompano, yellowtail, bonito, turtle from the factory ship nearby, and even sierra. That day, though, my eyes told me to eat slowly and smally, as if without hunger, taking every taste as a gift. Into the night we went, eating, swimming, eating, taking every offer from the good day. Ordinarily a solitary person, I swirled and let swirl the lives around me, into every kind of sustenance, into the fires of the night.

Next day I found the pelicans again. I watched them all day, did not go away from the thatched roof. Just before time for the coconuts, I watched a pelican dive close to the shore. It shot down on target but did not for a second disappear in the water. It flopped on its side and lay there. Farther out, other pelicans dived, waggled food down, and flew. The close one floated on its side. I jumped from the hammock, waded out to where it lay. With astonishment I realized it had made a dive in water not a foot deep. Its neck was broken. I had not expected that of pelicans.

24. When The Whale Came

The two rangers were standing on a bluff above the sea. They looked out on the ocean, but they were not just glancing around. They were staring. I was heading south to Santa Cruz, California, from Pescadero, which is about 50 miles down the coast from San Francisco. It is said that the sea is eternal, which would only mean that it tended to be boring. But the man and woman, standing by their ranger-green cars, were attending to something they found worthy of the look.

Indeed it was. A mystery. Just in from the horizon, there was a long, smooth mound. It was too regular for a rock not seen before. No submarine would tip so flexibly or draw so little water. It stayed there a long time, and I made my way south.

The next day, I returned to the spot, at Fiddlers Cove, which is just three miles from where I live. And there it was.

On the littoral, between high tide and low, lay the entire body of a blue whale. Now, I knew this before: The blue whale is the largest mammal ever to live on earth. Larger than a dinosaur. One sees the drawings of both together in textbooks and understands that. Not so. One understands nothing.

Standing on the bluff, perhaps thirty feet away and above the whale, I was astounded. It looked so small. And I just could not understand that. Small? It was so enormous that I had to turn my head to look at all of it. Small? My imagination, the endless deceiver, had me look for a mile-long beast able to capsize ships, gorge on people, rule the sea. But I had to go down to it.

Then I knew. It was that I could not diminish it enough to make it real. I walked it, touched it, looked up at that belly to the sky. I am two inches over six feet, and I looked up and could not see over it. I

148

measured. About 85 feet long. That means 90 to 120 tons, give or take the weight of 30 million pygmy shrews, the smallest mammal on earth. Now I got it, the bulk, the quantity. It would have taken crampons to see it all, to climb it all over. Perhaps this is the way: A semi truck and two trailers of maximum size are 20 feet shorter than and weigh half as much as the whale.

I wanted to understand it, to discern it, to assimilate it. It was here at my fingertips, an event in the real world without any reference points in experience. It did not even occur to me then to think of myth, scripture, fiction, or Greenpeace. Only I and it were thrown together on the beach.

Who could let it go? The blue whale stays 200 miles off the coast here, while the gray whale, a mere tad at 45 feet, comes close enough to see, in life. And the blue whale is the wrong whale, as opposed to what we still call the right whale. It is wrong because it sinks when it is harpooned or dies. The whaler has only to fetch and kill the right whale.

Captain Charles Melville Scammon—such an apt middle name—wondered at the blue even as he discovered the gray's place in the lagoon in Baja California now named after him. The captain noted a blue of 95 feet, with a jawbone over 20 feet long, the whole of which whale he thought weighed 150 tons.

Blues are now protected, but at one time after Svend Foyn invented the harpoon which blew the blue up inside but left it afloat, a female blue lived a feat Hercules would have envied. This blue, her vital statistics unknown, pulled a 90-foot whaler at five knots for eight hours. The whale had a 500-pound harpoon in her on a half-mile of four-inch rope, and the ship's engines were full astern. All that power in the gyre of her nether third, along with the flukes.

About the name, we have even got that wrong. This whale was once known as the sulphur-bottom whale. The color is about right, but it comes from diatoms which build up on the pearly underside, not from the whale's own color. And blue is even worse. The top is a steely slate, and so are the pleats along the midsection. These are like accordian pleats, for expansion and contraction, perhaps to help move along the ton or so of krill ingested through the baleens, the strainers hanging from the upper jaw. Someone once suggested that the pleats and the expansion made it possible for the blue to stop. One wonders why it would be necessary for it to stop. Ever.

This one stopped on the beach after turning slowly around and by rocks off the shore and coming right in here. I kept watch on her every day for all the month she was here. I could say I did it because I am a naturalist. But that would only be to find motives in my education. I went there for esthetic reasons not related to the works of man, the canons of my own species. And just because. I have trouble distinguishing among natural history, history, and personal history, so I get as esthetically excited about whales as I do about symphonies. I could explain that more, but it is really just the way I am at play in the fields of the earth. So I went there, often.

Others had other good reasons. The list of official people who wanted the whale grew daily. The feds wanted it, the universities at Santa Cruz and Berkeley wanted it, some county people claimed it. No one knew who "owned" the whale, which is a strange idea to begin with. And the newspapers got it, the television newsreaders got it, the public came in numbers estimated as high as 50,000. Somehow, the University of California at Santa Cruz firmed its claim, but no money to get the bones and put them back together was available. The Pentagon said that the use of helicopters belonging to the military was "inappropriate." Only to save or kill people, and they don't do whale slices.

So the people at Santa Cruz decided they would just do it themselves. And they did that. Almost. Day after day they came with knives and sharpeners and hay hooks and ropes and their own trucks. A short helicopter visit from Air Crane in Watsonville removed the skull. Students and volunteers came to carve, to get the bones and take them 35 miles south to Santa Cruz and put them together in order to learn and to teach. They could only work in daylight hours when there was a low tide. At high tides, the blue could move a bit, enough to pin a human leg under it. And one does not climb around with a sharp knife on a gigantic piece of blubber. Slowly, the bones were hauled up the beach.

Those bones are drying now at the university, out in a field. After that long process is done, there is to be a building. It will be odd to see it there, but no more than sitting with it every day.

Finally, it worked out that this county, San Mateo, had to get rid of the tons of blubber left when the bones were all gone. The crabs and fish and gulls and tides could have done it free, but the county hired a whale remover for three days at over $5,000 to haul the remains to a distant county dump. Our county dump at Pescadero had refused it, since the dump does not take dead animals.

One day I went there to Fiddlers Cove to see what there was. Some tissue pieces were there, the flukes were whole and on the beach. Perhaps a dozen vertebrae, shrouded in gray blubber, were there. But a few hundred yards offshore, a great patch of blubber crested and troughed with the waves. I had been there every day, sometimes for hours or more. I had heard the chatter of the watchers about looking for Jonah, and the comments about the smell, which was only like bubbling, homemade soap. Dawns and sunsets I had watched, without it being a visit, without sitting at a wake, without feeling the sadness so many people expressed so often. And I did not yet. I surprised myself in another way. I spoke to it. I said, Go on, keep going, go on back to sea. It felt like something I wanted to say.

Later in the day, the blubber came back to the same spot, even though the blue has extraordinary hearing. I told the students from the university, tying pieces they wanted to the rocks to save them from coming high tides, that I would ask the whale remover to leave them alone.

Then I went down the next morning. I looked at the beach. There was nothing there. Flukes, bone, tissue, everything was gone. No blubber. Not a sign. I looked up and down the beach. I was not ready so I pretended. I looked at the rock formations for fossils, I nudged the worn ice plant with my foot, I looked out to sea for a ship. And I sat on the bluff above nothing but sand and rocks and tide and no gulls. And the sadness came. For me, not for the whale.

The whale was gone.

The sanity of thinking of dawn redwoods and water fleas as part of one's family is a matter of point of view. The reality of it is beyond question. We—and otters, pelicans, whales and so on—are thrown here now together in the same home and made of the same materials, differing, as individual humans do, in the arrangement of the genes and other particles.

None of that reality makes good, objective, *a priori* reasons for attitudes which see other animals and the plants as having shared histories with one human. Nothing compels us to such attitudes. There are no given values except for human ones, chosen ones. And there is absolutely no empirical urge to like the world. Yet it happens. In *Self And World*, Eli Siegel wrote that "man's deepest desire, his largest desire, is to like the world on an honest and accurate basis," and that we lower other people or objects in order to have contempt for them only when we are unable to like the world.

Whether one agrees with that or not, it is a choice humans have and can make without any romanticism or sentimentality. It merely takes a certain education of the emotions rarely found in our schools or homes or any other part of our society. It may be that affection for such things as cities and boring jobs is difficult to feel but easy to find in needn't-running rabbits or daphnias, Enos Ralstons or waterfall keeping.

Obviously, it happened in the case of what the gifting birds left. When I arranged their gifts into perceptions and senses and ultimately a sense of having place and being animal, the arrangement, as ephemeral as it may be itself, included other animals and the plants of the earth. That much happened when the gifting birds came briefly for the roosting.

VIII

THE ROOSTING

Gifting birds in circles left places scattered where I found them. There were no labels or attachments or threads that led anywhere. Each one had a perception about it, an emphasis that might be drawn, but one which might fit with other stressed flavors. One place might make me think of some feeling of time, another of a person. Enos Ralston could have fit the human niche, but his time got me. The pelicans gave me a quick perception of movement I could make a human comparison to, but they as pelicans came harder to mind as creatures who can get caught in their own mistakes, just as we do.

I tried to call the gifting birds together for a roosting, and called again, but they would not come. It seemed obvious that I had to make the connections. If there were any threads of meaning, I had to make them. I had the perceptions, as anyone does, and I had to make sense of them, as anyone does. These might have been someone else's samples, but they were not.

So I browsed around in them, and wondered what they meant over a time, and took the strength of each one. And the sorting out I did made the only roosting I shall have. The quick, short, strong perceptions shone with their own refractory insistence, and I caught them as I could. Left by the gifting birds, the morsels were mine to arrange, if I could.

In the heron's feather and the *cabecita*; from the gratuitous echoes of Enos; and marked always with the cadence of silence which belongs to solitude, there was something which came as a sense of my own biological time. It seemed to be a completely unconscious moment, but one of unshakable knowing. Bald attention without words, the animal instant. Its first appearance was with time, the mute event, the first perceptions which went together as one part of a sense of place.

The sense of biological time is not the same as the measuring of movement. It is more the sense of having a present only, without past or future. Movement is of course the data of time, but it seemed obvious that in one individual biological creature, movement meant both more and less. It became translated into the markers of time, but more closely it meant a sense of where one's body is in the immediate space, a kinesthetic sense. And it referred to space out there, the moving continents, the mountain, the edge of the Arctic Sea and the mereness of tundra. Surely, that gives us part of our sense of place.

Other humans came along to give me perceptions of our place. Always and only, we respond as they come, one by one. But because we communicate one to the other, we also find them, lovingly or hatefully, in groups. People and the works of man form our human web. And so the places left by the gifting birds gave their quick perceptions of single beings like D. B. Wesson, or the places we make, like plastic planes and the cities of our lives. Being human animals, we probably notice other people more than anything which gives us a feel for place.

If that sense of other people is our most specific perception of place, the one we notice most, then the esthetic sense is the most inclusive. It works the pleasures of our senses and minds together in a bread we savor and want to break together. We point to, say, a sunset in order to let others in on it, and at least no one minds a sunset or wants to cut it down for profit. It is only ignorance which says that the hedonist cares solely about his own pleasure. There is nothing wrong with pleasure, and "own" is a misconception. The pure hedonist is eager for the pleasure of others; and the esthetics of place is the most exhaustive combination of the pleasures of the senses and the mind. Everything must be attended to, realized, related, fused, or separated, as the moment asks. In speaking earlier of Trigant Burrow and his concern for the mood of man and theories of attention, his words "cotention" and "ditention" came to mind as successful and unsuccessful attention. The overall feel of an esthetics of place is clearly cotentive, that quick feel at the back of the mind that one is where and when these animal instants occur. It is important, in sorting out what the gifting birds left, to put them at the front of the mind.

It is commonplace, for example, to point out that the earth is our home. Putting these perceptions together, however briefly they may roost, gives the earth inside another feel. It is not the romanticism of the soil between the toes, nor is it the wilderness urge given us by the city romantics like Thoreau and Emerson. It is more in the steps of John Muir, for whom the literal earth was home. Like it or not, walk as a way of life or not, this is in fact our home, and not a mere collection of countries.

To think the earth our home is to recognize and even attend to the other animals and the plants here. For our own survival, and for the sake of each organism in itself, we ought but do not yet have a sense of future debt. Nothing insists on such an attitude, for we shall, no matter what, probably join many other creatures in extinction. That means nothing in terms of an indifferent universe, but somehow it matters to me, to many of us. It is just a nostalgia for what is to come, as if a meliorist had empirical evidence that things would get better, and wanted to see if the footprints ahead were charming.

From the gifts of my birds, I put perceptions into senses and found a shelf space for each. I could see that they were my categories, defining what a sense of place is, and even so might not last out the night of a gathering like this. One takes such chances with words. And when I decided to read the shelves, see if my sorting had some space worthy of even a transient cote, I found a collection of perceptions and a sense missing.

By assuming it, I had left out the self. Such an obvious entity needs no attention. Had the gifting birds left anything, places strewn, from which I might find a few quick perceptions and fashion some sense of self? No, not so, for I, the agent, the perceiver, the responder, am simply here. There was something though.

Over and over in the gifts from my birds, I found one shard not given as a place from which perceptions and senses grew. I found it often; it came without recalling, as if I kept stumbling on it. And I kept speaking of it: the animal instant. Suddenly it seemed that what the gifting birds had left, the most valuable tidbit, was a sense of

myself as a biological creature. The prize was to be able to say that I
am an animal.

It has been said that man is the link between the apes and
human beings. Included in such a wry statement is an assumed
ranking: man higher than apes, human beings higher than man. And,
since man is the species which ranks things, we know who will come
out on top, given any kind of listing. We pick, for instance, complexity
of the nervous system as the essential factor, and man leads the list. We
do not use longevity of individuals as a yardstick because then the
lowly sea turtle, among others, would outrank man.

For centuries it has been our habit to claim that man is not an
animal. It is a common practice among religions of the world. Christ-
ianity, for example, tells us that only man has a soul, and there were
times when it said only white men had souls; it is, this soul, a human
contrivance. But the attitude is also in our daily speech and literature.
We refer to people we do not like as "just animals."

We also hear, from philosophers, anthropologists, poets, his-
torians, and other thoughtful people, that man is higher than or
outside of the animal niche on the planet because of his mind. Only
man can reason, only man can plan ahead, only man has civilization.
There is nothing very singular about that; those are the things man
does.

Others, in another ranking, say that man is not civilized, that he
is one of nature's mistakes. Man murders, is cruel, and destroys his
own habitat. Other animals do not do these things—or they are
extinct. Some naturalists and environmentalists say that animals lead
better lives and do not menace the earth, that the company of the
other animals is preferred over that of people. There are hermits and
recluses with that sense of the world.

We have put ourselves outside of—above or below—the rest of
life on earth. In much of the balance of that life, we own the scales but
think we are not one of the weights. It is not possible for man to stay
out of nature. We are not Hamlet that we can ask about being or not.
We are already in it. The question is about continuing. There is no
alarm to that. Whether man survives or not does not matter at all.
The universe will go on spinning and rhyming. But I am human, and I
like the world; and that kind of hedonism, like our shared histories,
ought to be passed around.

One thing I know to begin with is that letting your imagination browse around in man the animal is not a way to cure everything. Or anything. We shall still cut too many trees for us to keep having them for very long, we shall still filch the earth's largesse in the quest of profit without a tomorrow.

Much of the objection to considering man as an animal is that it is not exalted enough. Look at the glories of man—his mind, his civilization, his technical accomplishments. He is able to rise above anything, do anything. A special being. Of course that is man's ranking of himself. Thought of as a natural being, man is not remarkable. What he does is simply what he is like as a species, neither good nor bad, save in his own eyes. Every species is, after all, like every other, like some others, and like no others.

So none of it mattters at all. Except a feeling under the skin, in the bones. There are no given facts telling me this, coming from the world. I can lie on a beach and feel that I am turning away from the sun, not it from me. I can try to understand why otters are worthy just because they are cuddly. I can hope to know what the difference is, in war among humans, between killing civilians and killing soldiers. I can look, in a moment without past or future, at a great blue heron friend suddenly gone on its way, leaving me a short, bereft silence. These are points of view, ones I think I cannot understand without knowing I am an animal, an individual of the animal man. There are certain threads, both broken and continuous, which give me place as geography, as people. But the gifting birds left more. There is an "out there" and an "in here." I think the gifting birds have told me that I cannot be fully human until I feel I am an animal.

It makes a connection from out there to in here, one which exists whether I like it or want it or not. As man and animal, that feeling reaches into my blood and out to the sea, both with shared salts and mysteries. Sitting in the woods at night alone, strange sounds come to me, and they go out from me when I walk. Out to rabbits who need not run, owls who need not fly. I look at the cool blue of azurite and wonder where the copper is in my own body, for I know it is there. Looking at other animals, I recognize my face in theirs, their dreams in mine. We know each other only one at a time, but we share the visions of unmet minds. These feelings of connections, dim and unnamed, tie me to the earth, to what is in it and on it, and I have no way of choosing anything anywhere to leave out. I am an animal made of earth—the people of it, the seas on it, the metals in it.

This attitude of mine that I am an animal, that man is animal—that such a condition is acceptable—is not exactly popular. In the articulate mind of man, "just" an animal breeds "just" a slave, merely a plant. Redwoods, clay, eagle, man, oak—what exactly is the reason for making one "higher" than the other? It is only that it seems to me now that an attitude which ranks its habitat or wants a divorce from it is dabbling in fatality.

Which, after all and once again, does not matter. Yet, caught in a fog, making love, surprised by a sunrise or earth set, listening to music, peering at the microscopic world of a drop of pond water, that tingle from scalp to toe tells me it all matters. To me. That is enough. I know that my home is as much miasma as meadow. The feeling embedded in me informs me that I am by chance a life among lives, thrown here not there, now not then.

So, without choice, still I choose to follow my bones, among the animals of the earth.

THE GIFTING BIRDS

Designed by Richard Firmage, Salt Lake City
Composed by Alphabet Soup, Salt Lake City
Printed by BookCrafters, Chelsea, Michigan
Bound by BookCrafters, Chelsea, Michigan
Typeset in Garamond and printed on acid-free paper.